the whites of
their eyes

Also by Roger Ford

The Grim Reaper:
The Machine-Gun and Machine Gunners

the whites of their eyes

Close-Quarter Combat

Roger Ford & Tim Ripley

Chechnya material researched and
provided by Charles Blandy

Brassey's, Inc.
Washington, D.C.

First published in the United States of America in 2001 by
Brassey's, Inc.
22841 Quicksilver Drive
Dulles, Virginia 20166

ISBN 1-57488-379-8

First published in Great Britain 1997 by Sidgwick & Jackson,
an imprint of Macmillan Publishers Ltd.

(alk. paper)

First Edition

10 9 8 7 6 5 4 3 2 1

Contents

Introduction

'Don't fire until you see the whites of their eyes' is a legendary infantry fighting order. It conjures up images of combat at its most basic, where every man could smell or touch his opponent, and feel the life drain away from him as he died.

In this century, the development of hi-tech weapons of mass destruction, such as the long-range bomber, ballistic missiles, nerve gas and the atomic bomb, seemed to depersonalize warfare. Whole countries could be obliterated at the touch of a button, and there seemed little need for the lowly foot soldier, or so Hollywood imagined. Generations of infantrymen, though, have known a different story.

From the bloody slaughter on the Western Front in the First World War to desert battles in the 1991 Gulf War, victory has gone to the army whose infantry were able to fight through the enemy's defences and evict him at bayonet point from his positions.

The weapons and tactics of the infantryman have undergone dramatic changes over the past hundred years. No longer is the infantryman seen as expendable cannon-fodder. He is a well-equipped, armed and armoured (with lightweight body armour) specialist in his own right.

Thanks to light armour-piercing rockets, infantry can take on enemy tanks and blast the opposition's bunkers at long range. High explosive, fragmentation, stun and smoke grenades mean the infantryman can clear out the occupants of field defences or buildings. The development of light automatic machine-guns and assault rifles has dramatically multiplied the firepower available to an infantry section. A modern British infantry section of eight men armed with LA85/86 automatic rifles/light support weapons can fire off almost fifty times as many rounds a minute compared to its 1914 equivalent armed with 1907 Short, Magazine, Lee-Enfield bolt-action rifles. In addition, small night vision devices mean the modern infantryman can fight at night and in all weathers. No longer can the enemy expect a natural pause in the battle when darkness falls.

With lightweight Kevlar body armour and helmets, the infantry-man now has an unprecedented degree of protection. Although not a panacea, they protect the vital organs from bullets and shell splinters, making it more likely that the wearer will survive an injury. Closing and killing the enemy is still the main role of the infantry, and he is now provided with a deadly array of bayonets and killing blades to allow him to do the business if necessary.

The infantryman is now part of a combined arms battle group, with close support provided by tanks, artillery, mortars and aircraft. To allow him to call in heavy firepower every infantryman must be able to use a radio.

Infantry tactics in the 1990s are unrecognizable from those employed in the early months of the First World War. In 1914 the infantryman only really had two weapons – a bolt-action rifle and a bayonet. There were almost no machine-guns or mortars attached to infantry battalions, few grenades and no steel helmets for protection from artillery shell shrapnel. The infantry could only take ground by evicting the enemy at bayonet point, and defence was conducted by massed rifle fire. The inevitable result was massed bayonet charges that invariably ended up with heavy losses and little ground taken.

Within a few months of the First World War starting, however, British, French and German infantry were all provided with more effective weapons and equipment. They also became the masters of defensive warfare, building huge trench lines, deep bunkers and massive barbed-wire entanglements. Rapid-firing machine-guns and highly accurate sniper rifles, backed up by mortar and artillery fire directed by observers using field telephones to pass orders, made it suicidal for anyone to venture out of the trenches into no man's land.

By 1917, the Germans and British had developed new infantry tactics to break this deadlock. Light artillery was attached to infantry units to allow them to blast strong points, light machine-guns were provided to enable the infantry to carry their firepower forward, and frontal attacks on the enemy were spurned in favour of attacks against weak spots. The British added tanks to these tactics to crush the barbed wire entanglements.

In the Second World War these tactics had been adopted by all Western armies, and the infantryman was no longer relegated to purely defensive actions. Russian, Japanese and Chinese armies were less advanced than their Western counterparts, and still relied heavily on the massed bayonet charge as their main infantry offensive tactic. In extreme jungle and winter conditions, against poorly supplied and led troops, it usually carried the day, as the Japanese found in their 1941–42 offensives. When they came up against determined and well-prepared defenders, though, machine-guns, artillery and tanks mowed down these 'human-wave' attacks with ease.

Since the end of that war there have been hundreds of conflicts. These have ranged from large-scale conventional wars, such as in Korea, Vietnam, the Arab–Israeli Wars and the 1991 Gulf War, to small counter-insurgency campaigns. All these wars required hundreds of thousands of infantrymen to fight in the traditional way, both holding and taking ground. And as well as carrying out their traditional combat roles, the infantryman has been called upon to carry out unconventional tasks such as policing United Nations cease-fire lines, winning hearts and minds in counter-insurgency operations or holding riot lines during civil disturbances.

The rise of international terrorism since the 1960s, with its brutal attacks on civilians in aircraft hijackings, embassy hostage sieges, random bombings and the like, has forced many governments to turn to their armies to field highly specialized units which can rapidly deploy to take on terrorist groups in close-quarter combat. These battles are swift, violent and bloody, requiring highly trained and motivated infantry soldiers. Counter-terrorist tactics grew out of traditional house-to-house combat skills, and the units that specialize in them are counted as some of the world's best combat soldiers.

Counter-terrorist soldiers need to be highly proficient killing machines. Even the slightest mistake during a rescue mission can result in dozens of innocent civilians being killed and injured. Speed, aggression and precision are essential skills.

While the infantryman's weapons and equipment have changed dramatically over the past eighty years, many things in the life of the infantryman of the 1990s would be instantly familiar to his predecessors from the Great War. The infantryman often has to carry all his weapons and personal equipment for days on end, in all weathers. He has to endure cold, hunger, lack of sleep and shelter and still be fit to fight.

In the First World War the phrase 'poor bloody infantry' was coined to describe the condition of the combat soldiers who manned the trenches. Front-line soldiers in Korea, Vietnam, Afghanistan, the Gulf, Chechnya and Bosnia have all had to endure terrible conditions. Mud, rain, non-existent rations, polluted water, heat, tropical diseases, all had to be coped with, even before the enemy had started inflicting casualties.

Other things that do not change are the reasons why men keep fighting in such conditions. Much is made by politicians, religious leaders and generals of just causes, religious rights, patriotism, fighting for freedom and justice. The nearer you get to the front line the less convincing such reasons become. Combat infantrymen, whatever army they come from, are locked in a grim battle for survival. While soldiers may sign up for a variety of reasons – conscription, idealism or to become career soldiers – once the bullets start flying self-preservation takes over. Well-led and trained soldiers will react positively to combat, bonding into an effective team that will advance under fire to defeat the enemy. Soldiers will see their salvation in working together and protecting their comrades. Many nations with professional armies see unit cohesion developed best by establishing regiments or battalions that work and train together for long periods of time during peace.

Poor leadership and battle preparation will result in units cracking under fire. Then individuals will only think of themselves. They will not attack, and as soon as they are personally threatened they will run for safety in the rear or surrender in the hope of getting better food and conditions. Some armies have tried to compensate for the poor training and motivation of their soldiers by imposing harsh discipline and reprisals for anyone who shows a lack of moral fibre in the face of the enemy. The Japanese, Soviets, Red Chinese and Iranians, for example, have all used these methods. They could only get away with fighting in such a way because of the total disregard they showed for their soldiers' lives. Life was cheap for them, their soldiers knew they were expendable. If the enemy didn't kill them, then their own side would.

But whatever the reasons responsible for bringing infantry on to the battlefield, once close-quarter combat begins all soldiers experience the same rush of emotions – exhilaration, fear and stress. And however sophisticated their weapons and equipment, when they see the whites of the enemy's eyes, as the following chapters will show, the action is often nasty, brutish and short.

Chapter One

Fifty-Two Months in Hell

Offensive à Outrance

During the last third of the nineteenth century, a new order seemed to take control of warfare: technology. The 'new' factor manifested itself in many ways, but the most significant and far-reaching changes it wrought, in fields such as chemistry, metallurgy and production techniques, combined in one area in particular: projectile weapons, which had already, even in their most primitive form, come to dominate organized combat, and which now took over completely.

In the period between the American Civil War and second Boer War, at the very end of the century, one simple invention – the drawn brass cartridge, with encapsulated detonator and projectile – made two new types of infantry weapon practicable: the repeating rifle and the machine-gun. Between them, these two weapons lengthened the infantry battlefield to the point where adversaries saw each other as not much more than dots in the distance, and when we add in a further simple factor – the introduction of barbed wire to hamper the advancing formations – we can see quite readily that by the beginning of the twentieth century the nature of the battlefield had changed out of all recognition in less than a lifetime. Large-scale close-quarter battle was, it seemed, a thing of the past, and future infantry battles would be fought at a range of some many hundreds of metres. In the event, such prophecies were well wide of the mark, as we shall see, and it was to be close-quarter infantry fighting which eventually proved to be the deciding feature of combat on the battlefield after all.

If we examine the character of the armies which turned out to fight the Great War in the summer of 1914, we can make a number of fairly straightforward conclusions about the nature of the war they thought they were going to. That may sound self-evident, but, in fact, such an examination is very revealing, for it demonstrates that the strategists on neither side had grasped the fact that technology had already taken over from valour and skill-at-arms as the deciding factor. Thus, in the armies of both sides, we see cavalry forming up beside infantry units equipped with machine-guns, even though every serious theoretical study – and a handful of useful, practical examples from almost the previous half-century, since the short-lived Franco-Prussian War of 1870 – showed that in the presence of the latter, the former was doomed.

The clearest sign that there was no general understanding that the real rules of war had changed comes from the strategy under which the French planned to fight. They envisaged a war of sweeping movement and short, decisive battles, not at all unlike those that Napoleon had fought at the beginning of the previous century, where the side which best demonstrated its *élan* and will to win would indeed carry the day, and probably at the end of its bayonet. *Offensive à outrance* was the central theme of Plan 17, the French blueprint for war against Germany (it was so called because it was the seventeenth of its kind since the end of the war with Prussia in 1871), and Plan 17 allowed for no digging-in in defence, no deviation from the principle of sending the troops forward to meet the enemy and shatter his attacking formations upon their own, and demonstrated no up-to-date military thinking – or even common sense – whatsoever. The men who formulated the French Army's battle plans were clearly living in a fantasy world. They calculated that a line of infantrymen could cover 50 metres in a dash of twenty seconds before meeting the enemy with the bayonet, and that defending enemy riflemen would take that long to bring their weapons up, take aim and fire. Once. There are few polite terms to describe the average soldier's view of such an appreciation (especially the 'average soldier' forced to undergo the reality of it), and if there are, it is unlikely that many average soldiers are familiar with them, but none the less, such was the French adherence to the doctrine of *offensive à outrance* in the period leading up to the Great War that there was no argument put forward at regimental level when this ludicrous assumption was put into practice. What should have been a murderous charge with levelled bayonets inevitably became a bloody rout, time after time, as the flower of the French

Army was thrown away (almost a third of all French Great War casualties occurred in the first three months of the war) in mindless assaults which might – but only *might* – have been successful at Waterloo. Almost incredibly, there were those on the French General Staff – and they were influential, sometimes central, members such as Robert Nivelle, who briefly and almost catastrophically replaced Joffre as commander-in-chief, was one of them – who continued to advocate the same tactics even in 1917.

There was no alternative: Plan 17 was the blueprint for victory; Plan 17 was the spirit and the pride of France; Plan 17 relied upon offensive action now; Plan 17 required infantrymen to assault prepared positions with no protection save for the presence on the battlefield of quick-firing 75mm guns loaded with shrapnel (another ploy which might have worked at Waterloo, but which had been shown to be ineffective a decade earlier in Manchuria); Plan 17 was the death warrant of hundreds of thousands of Frenchmen.

Thus did a modern army waste itself in trying to bring a mechanized enemy to close-quarter battle. In the process, it became clear – eventually – that the sort of tactics which had worked as recently as 1870, when cavalry and infantry could still charge an enemy position with a chance of success (though not with any real measure of security) and subdue the enemy in hand-to-hand fighting, were useless against repeating weapons, let alone *automatic* weapons. The dead hand of trench warfare was poised over the two sides, and when it fell it smothered *élan* once and for all.

Not that the British Army's tactical doctrine was any better, or even altogether different; the 1909 Field Service Regulations declared that infantrymen would suppress the enemy's fire as much as possible with their own as they advanced, using fieldcraft (such as it was; it consisted chiefly of fire-and-movement and short rushes) to protect themselves as much as possible, until gradually a firing line was established about 200 metres from the enemy's position. Then, once the firefight which followed had been won (with the assistance of the artillery, firing field guns over open sights from positions on the battlefield itself) and enemy fire slackened, the attackers would charge forward with the bayonet and finish the job. (This was actually just the second movement of a set piece in three acts: Act One was the artillery's preparatory (shrapnel) barrage; Act Three was the cavalry, unleashed to chase and kill the defeated enemy.) 'Opinions varied widely about just how much stress should be placed on fire and how much on the bayonet,' says one

9

knowledgeable commentator, Paddy Griffiths, 'as also upon the precise interval between men at each successive stage of the operation. Some authorities envisaged a very tightly packed line charging at an enemy who had not been fully suppressed at all, on the basis that a battle which consisted only of stand-off fire could never reach a decision.'[1] One is constrained to wonder if the formulators of such tactics had learned anything at all from the events of the previous decade – or even if they had *heard* of them. Certainly, the tactical plan didn't differentiate between enemy troops in the open and enemy troops in an entrenched defensive position. In sum, what the tactical doctrine best demonstrates is a callous and cruel disregard for the life of the 'poor bloody infantry-man' – something which was largely to disappear as the war progressed, but which seems to have reared its ugly head again half a century later in Vietnam.

At the start of the Great War, British infantry tactics were still rested in the out-of-date doctrines of the nineteenth century. The infantryman's main weapon was a bolt-action 1907 Short, Maga-zine, Lee-Enfield, Mk III rifle, the famous SMLE, with a bayonet. Steel helmets had yet to be invented and machine-guns had not been issued in large numbers.

The British Army of 1914 was a small, professional force manned by regular soldiers, many of whom had seen action in Britain's colonial wars. Khaki uniforms had been introduced for the Boer War at the turn of the century, but many serving soldiers could still remember going into action in colonial conflicts wearing the famous British Red Coat. It was not until early in 1915 that every British soldier had been issued with a steel helmet to provide protection from artillery shrapnel.

Much as in other armies of the period, its infantry tactics centred on controlled massed volleys of rifle fire. Infantry companies and battalions formed up in long lines in the open to give soldiers and their commanders good views of the enemy. Battalion and company officers would then shout orders to their troops to fire in volleys. The professional British infantrymen prided themselves on being able to fire off fifteen rounds a minute from their Lee-Enfields. The idea was to create a 'cone of fire' in the front of the regiment in which some bullets were guaranteed to hit something. Each com-pany was ordered to aim at a different height to ensure that the bullets spread properly. While it was recognized that close forma-tions were not tactically sound on modern battlefields, leading to the adoption of loose-order skirmish formations, with men standing two to three metres apart, the troops were still kept under close control by officers and NCOs. Only senior officers were allowed to control fire.

The bayonet was still seen as the main assault weapon. Battalions were to form up in line some 1,500 metres from the enemy, and then steadily advance until they were within rifle range of the opposition, usually about 500 metres from them. Then the battalion was to rush forward and assault the enemy. The accepted theory was for assaulting troops to ignore mounting casualties and press on forward. As the assault troops got closer, the enemy's morale would falter and his weight of fire would lessen, or so the high command said.

During the initial battles of the First World War these tactics proved to be the undoing of the British, French and German armies. Casualties were horrendous. Regiments fell apart under the weight of massed rifle and machine-gun fire. Artillery proved more deadly than anyone had imagined, and troops in the open suffered horrendous injuries due to the lack of steel helmets. Troops in cover proved impossible to dislodge. Large formations of infantry could not close on the enemy without suffering massive casualties. The bayonet charge did not scare the enemy, as they simply held their ground and shot the enemy formation to pieces. The deadlock of trench warfare had begun.

Stalemate

The British Army received orders to begin entrenching on 15 September 1914. The lines of roughly parallel fighting, support and reserve trenches – broken by frequent traverses into sharp zigzags or crenellations to minimize the damage an assaulting enemy's enfilade fire (or shells and bombs) could inflict at any one point – were linked together by a system of minor communications trenches and backed up by approach and supply trenches which reached far into the rear echelons. They were studded with protective dugouts ranging in size from mere scrapes to quite large subterranean rooms, which contained forward command posts, communications posts, first aid posts and the like, as well as sleeping accommodation. The trench line (the term 'line' implies a feature with length but no breadth; this was hardly true. On the German side, the 'line' was six

11

to eight kilometres deep by the summer of 1916) stretched across the soaking levels of Flanders, through the industrial Artois and the clay and chalk plains of Picardy, as far south as Soissons, barely a hundred kilometres north-east of Paris, before turning east towards Reims and the impregnable fortress of Verdun and then south again, on through the much more difficult terrain of the Lorraine to the Vosges mountains, Alsace and the border with Switzerland.

For most of their length, the opposing front lines were separated by wide tracts of 'no man's land': uninhabitable terrain swept by the artillery, and by rifle and machine-gun fire, covered in a rash of shell craters and broken and bordered by almost-impenetrable tangles and thickets of rusting barbed wire. The nature of the terrain dictated its width, anything over 250 metres being considered enough to form an adequate buffer zone. Occasionally, however, no man's land would be very narrow indeed, perhaps no more than the length of a tennis court wide, and then life for the occupants of the trenches on each side became a tense struggle, with the enemy just a stone's – or, more likely, a bomb's – throw away.

The trenches were never 'home' for very long; under normal circumstances a battalion would be moved up to 'the line', as it was soon known, for up to a week (but sometimes for as little as three to four days, depending on the level of activity in the sector), and then be relieved to rotate through the support line, brigade reserve and a rest area. A battalion would be expected to hold a front perhaps a thousand metres wide, in a quiet sector, reducing to as little as a few hundred metres in one which was particularly active. For most of the time, the only significant enemy to be found there were the living conditions and the boredom. The offensives – the 'shows', the 'big pushes' – were relatively few, and in between them there was little to be done, and certainly no point whatsoever in provoking the other side into a senseless fire fight which would both cost lives needlessly and expend the stocks of ready ammunition which had been so laboriously man hauled from distribution points far to the rear. What little offensive activity there was was concentrated in the 'morning hate' (a fairly pointless exercise in itself, but one which, it was felt, gave men the chance to vent their frustration and ease their pent-up feelings), in desultory sniping, in patrolling and maintenance of the wire which was the real first line of defence, and sometimes in raiding the enemy's trenches, either with the objective of taking prisoners for the sake of whatever meagre intelligence they might possess, or simply as exercises in belligerency.

Patrols either had reconnaissance as their objective, or were designed to maintain dominion over a particular sector of no man's land and disrupt the activities of enemy working parties, generally those trying to repair breaches in the wire. The former, in particular, relied on stealth and a special sort of courage – the courage to creep across no man's land and through protective screens of barbed wire in order to set up an observation- or listening-post. The priority here was to remain undetected, and so reconnaissance patrols were made up of as few men as possible. Fighting patrols, on the other hand, might be as strong as a platoon of thirty to forty men (though they more usually amounted to a section). In the British Army, it was considered essential (for reasons of morale) that a fighting patrol or a raid be led by an officer (invariably a subaltern, for anyone of the rank of captain or higher – that is, a company commander or above – was actively discouraged, if not actually forbidden, from taking part, owing to the difficulty of replacing experienced officers). Captain Lionel Crouch of the Buckinghamshire Battalion, who revealed himself as having a light-hearted approach to the business of making war – in his letters home, anyway – described a fighting patrol's 'stunt'; it is necessary to read fairly closely between the lines to detect any sign of the extreme danger any operation like this entailed:

Combie, who is my second in command, took out a large patrol to round up some Boches who were suspected to be occupying some poplars and a sunken road about 700 yards from our, and 100 yards from their trenches. Smaller patrols of my Company had found them there two nights in succession. His scheme was to bomb them out of their post at the end of the poplars and drive them up the road into the arms of another party. Lance-Sergeant Baldwin put four bombs [grenades] into the Boche post; these were followed by loud shouts and groans. The Boches manned the whole line of poplars and opened fire and threw bombs. About twelve of them charged along the road towards our own trenches. Lance-Corporal Colbrook stood in the middle of the road shouting 'Hands up!' The Boche was shouting 'Deutscher, Deutscher!', evidently taking our people for Boches. Colbrook would have been mopped up but Corporal Baldwin and Goldswain each put a bomb into the middle of them, knocking out all but three. These swerved round Colbrook and began firing from the hip. One charged on to Goldswain, who fired his rifle from the 'on guard' position when the Boche was practically on his bayonet. A Boche bomb then burst behind Goldswain, throwing him down nearly on top of the dead Boche. Another was also shot and the remaining one fled in another direction. Heavy Boche reinforcements came charging down the road and nearly bagged Corporal Baldwin and Co., who escaped by the skin of their teeth. The patrol was then reorganized, and Combie made a counter-attack, driving the Boches out of their listening-post again, but the Boche reinforcements occupied the line of the poplars and opened fire and threw bombs. We have three men very slightly wounded in the face by bits of bombs, including Alan Crouch and Goldswain. The patrol retired in fine order, halting occasionally and facing about in correct style! Combie reckons he put out from twelve to fifteen Boches in all, killed and wounded. Anyhow, it was a very successful little scrap.[2]

Trench raids were more formal affairs; they were often planned and mounted at battalion or company level, and in company (that is, perhaps 100–120 men) or platoon (of 30–40 men) strength. Sometimes, however, they were undertaken by much larger detachments; one particularly well-documented operation, dating from late in 1917, when the technique was well established, involved men from two battalions of the Durham Light Infantry in a raid on a stretch of what the British called Narrow Trench, near the village of Chérisy, some fifteen kilometres south-east of Arras. The Chérisy raid was unusually large and elaborate and had been planned at divisional headquarters, with considerable artillery support at that and corps level, as well as the participation of aircraft of the Royal Flying Corps and of a detachment of Royal Engineers. Also included was a company of the euphemistically named Special Brigade, the chemical warfare experts, who would launch a localized poison gas attack. But this was only one of nineteen significant trench raids undertaken by the British Third Army during the month of September alone.

The Chérisy raid stands out from the rest by reason not just of its size, but also of its studied brutality. It involved no fewer than three separate phases: an attack by three companies of the 8th Battalion, followed by a second attack by one company from the 9th aimed at killing surviving Germans engaged in trying to reinstate their defences. As if that was not enough, that night the

Special Brigade 'engineers' fired 522 29kg bombs, each containing 13kg of liquid phosgene, a highly toxic respiratory agent. The British sat for 10 minutes, waiting for the German stretcher-bearers to arrive, and then, for a further ten minutes, poured machine-gun, mortar, gun and howitzer fire into the gas cloud.

More commonly, trench raids were simpler affairs, though still planned down to the last detail, and frequently mounted with the benefit of a localized artillery barrage to soften up the opposition and drive them out of the trenches and into the protection of their dugouts and bunkers. Their objectives were manifold, but typically included taking prisoners, so that the identity, readiness and level of morale of the units occupying the position in question could be established; destroying strong points and dugouts; destroying or capturing mortars and machine-guns, and killing as many of the enemy as possible. Raiding relied on surprise and aggression on the part of the raiders, using a variety of expedient weapons (many of them home-made, or suitably embellished) such as nailed truncheons, spiked iron clubs, sharpened trenching tools, billhooks, hatchets, tomahawks and heavy knuckle-dusters. Rifles, over a metre long even before the 45cm bayonet was fitted, were reckoned by some to be more of a hindrance than a help, and were often left behind, pistols occasionally taking their place. Hand grenades, either factory-made Mills No 36 bombs (introduced in imperfect form in late 1915 and being produced – and used – at the rate of 1.4 million per *week* by the end of the following year, by which time they were much more reliable; the hand-thrown version – there was one which could be fired from a rifle, too – weighed 623g and contained 85g of ammatol (a mixture of ammonium nitrate and TNT – trinitrotoluene) or ammonal (a mixture of ammonium nitrate, powdered aluminium and charcoal). Some thought it the close-quarter weapon *par excellence*, others were by no means so sure. Rather more precarious home-made devices, jam tins filled with explosives (usually lyddite, derived from the charges from unexploded artillery shells) and scrap metal, were ubiquitous.

J.C. Dunn (see page 18) tells us that up to early in 1916, the 'immemorial notion of a raid [was] to enter the enemy's lines by stealth or surprise, to kill or capture, plunder or destroy, and get away with no fighting or little'.[3] The raiders' tactics for these less-than-sophisticated operations were understandably very basic: arrive in the enemy positions without warning, cause as much damage, mayhem and confusion as possible, and then slip back out, perhaps

15

with a prisoner or two, roughly bound and gagged, but probably senseless anyway, to make the arduous return journey across no man's land, this time probably under more or less effective fire, if not from the segment of the enemy line which had been under attack, then almost certainly from neighbouring positions, trying to stick to marked routes through the entanglements and avoid tripwires and the like before exchanging passwords with the bloke on sentry-go, tumbling back over the parapet and scurrying away to the support trenches for a cigarette, a 'wet' of tea and the generous tot of ferocious Navy rum which was often the only quasi-official reward they received.

Sometime in March or April of that year, however, the Canadians – always innovative – recast the raid as a sort of 'battle in miniature with all the preliminaries and accompaniments, often manyfold'.[4] The ultimate destructive objective was the same, though. Captain Henry Dundas of the 1st Battalion, Scots Guards, described the arrangements for such a raid. Though the account is written in the present tense, it was not published – in Dundas's *Memoirs* – until 1921; Dundas was killed in action on 27 September 1918:

> We have raids almost nightly – fifty men and a couple of officers. Artillery preparation for about an hour on a fairly wide front so as to keep the Germans in the dark as to where the actual entry is going to be made into the trenches, then they ring off for five minutes; the raiders rush across, and the Artillery lengthens the range a bit and forms a barrage [an example of the proper use of a word which soon became distorted; it was borrowed from the French, and means 'dam'] behind the sector being raided. The raiders are generally over for about half an hour, and at a given signal are supposed to leap out of the trench and return with as much plunder, human and otherwise, as they can get.[5]

Not every raid went well, by any means – that the activity was fraught with danger is obvious, but in addition, there were many unknowns to be considered: timing (and not just of far-off support arms' contributions, but of the unit's own components), equipment failure and the presence of uncut (and hence, often insurmountable) wire were probably the most significant of them. The 2nd Battalion, Royal Welch Fusiliers, in the line at Cuinchy, just to the east of Béthune, on 25 April 1916, suffered accordingly:

Another raid has been planned by the CO and Rochford Boyd [a Lieutenant-Colonel; the liaison officer with the Divisional Artillery]. Fifty-five of B and C Companies are to go for the re-entrants of a small salient on the left of the road. Stanway was in charge in front. The night was dark, the air clear and still: the surface was dry: the whole front was quiet. Suddenly, at 10 o'clock, many guns opened up on the salient. They pounded it for fifteen minutes; then they boxed it [see below], pounding its surroundings until it was all over. Things went wrong from the start. 'Uncle' sent his contingent up tail first, and so late that they barely got out in time. Then most of them followed Sergeant Joe Williams, who made off half-right, shouting 'Lead on, B Company.' They ran on to uncut wire, were enfiladed by a machine-gun and Joe was killed. The gun was to have been kept quiet by the two new trench-mortars, the Stokes, detailed to protect the right flank, but both broke down when they began to fire. The few of B who followed their officers got into the enemy's trench, but found it empty. C Company's 2 officers and 25 men also got in, but all they could bring away was an anti-gas apparatus. Both parties were heavily strafed from behind. No identification was found, but a second sergeant was left behind dead. All four officers were wounded, one of them was able to remain at duty. Our other casualties are 21, so Fritz had the better of the exchange, unless our bombardment caught him.

Watching the raid from the fire-step was a hair-raising affair at the start, for our field guns had so little clearance that the draught of their shells could be felt on the back of one's neck. What was to be seen was like nothing else. Against the night there was a wild dance of red fan-shaped spurts of fire seen through a thickening haze. Soon, so it seemed, Stanway said quietly, 'The boys will be over now, they'll want some light, and he fired a Verey [sic] cartridge. It fizzled up and flared. Myriads of drifting specks of smoke and dust distorted its light. Through the translucent mist there loomed a figure running hoppetty-kick, he was a sergeant wounded in the foot. Then the light went out. In the brilliance of a second rocket two men crouched and shuffled, supporting a third. Two other men dashed over, one of them holding an arm. The strangely refracted radiance made them look unearthly. And all the time the unchecked machine-gun on the right stuttered in fits and sparked viciously. This weird, fascinating scene was played to the crashing notes of guns and bursting shells and bombs. After the briefest diminuendo there was quiet, and only the dark in front. Stepping down, I asked, 'What's happened?' Someone said, 'They're all back.'[6]

This account is fairly obviously the work of two individuals, the second part distinctly more literate and dramatic than the first, and describing the action from a different point of view. The 2nd Battalion, the Royal Welch Fusiliers (RWF) was unusual in that it counted two 'men of letters' – Siegfried Sassoon and Robert Graves – in its number, as well as Private Frank Richards, one of the few enlisted men of the First World War to write his memoirs (*Old Soldiers Never Die*, published in 1933 and widely acclaimed then and since). Perhaps more importantly, the battalion had, as its medical officer, Captain James Churchill Dunn, compiler of the remarkable *The War the Infantry Knew*, first published, in a limited edition of 500 copies, in 1938, and an essential source-book for any student of the history of the First World War on the Western Front as told by its participants. It covers, in journal form, the 2nd Battalion RWF's service in France from 22 August 1914 to early June 1919.

Contemporary accounts as colourful as those found in Dunn's compilation, even as heavily sanitized as they frequently are, are comparatively rare; more detailed – and particularly, more down to earth – descriptions, especially of the traumatic business of close-quarter fighting, are few and far between, for quite obvious reasons: most contemporary accounts were written as letters home, and few soldiers wanted to alarm their loved ones with too-graphic descriptions of life in the front line and the hazardous nature of an infantryman's duties (and it is unlikely that such descriptions would have passed the censors anyway). Most unexpurgated accounts of individual soldiers' exploits were written some time after the event (in many cases, quite a long time after the event) and sometimes lack detail and urgency as a result, though the work of Lyn Macdonald (*Somme* and *They called it Passchendaele*) and Martin Middlebrook (*The First Day on the Somme*) stands out.

An occasional addition to the artillery barrage which was the precursor to a heavy raid or set-piece attack, was the explosion of a subterranean charge, placed beneath the enemy's position by miners at considerable cost. Both sides used this technique, and the miners themselves occasionally met below ground, when horrific fighting ensued in the dark confines of the tunnels, usually with knives, pistols and grenades for weapons, but sometimes with nothing but shovels and pickaxes.

John Westacott was a civil engineer by profession. By the summer of 1916, serving with the 19th Canadian Battalion as an infantry-man, he had already been in action in France for sixteen months, and had fought at Ypres, where he was gassed, at Loos, and at St Eloi – where he had served with such conspicuous gallantry that he was commissioned in the field. He then transferred to the 2nd Canadian Tunnelling Company and moved to the north of the Messines salient, where he needed all the good luck for which he was becoming known.

On 30 May, while Westacott was away from the front on a short training course to learn about the Proto self-contained breathing apparatus, the Germans launched a surprise assault on the Canadian position at Mount Sorrel, taking not only the trench system but also the system of mines beneath and killing all the tunnelling officers present. Local counter-attacks won back a small amount of territory over the course of the next week, and on 13 June an offensive in force regained both the trench system and the entrance to the mines below. Westacott, at the head of a twenty-five-strong band of volunteer miners, had the task of securing the subterranean system, and carried it out handsomely despite being confronted in the process with the crushed and rat-eaten corpse of a close friend, which literally came apart in his hands when he tried to recover it. There remained a major problem, however – having had the run of the tunnels for two weeks, the Germans knew their geography well enough to attack them whenever they wished from saps of their own; so the Canadians doubled and redoubled their listening watches, alert for the first sign of trouble.

They had only a short time to wait, and when trouble came it was on two fronts at once, for the German engineers had dug out from their sap-head in two branches, threatening the Canadian system at two locations. Feverishly, Westacott ordered holes bored with hand augers, and called up a pair of 'torpedoes' – pipes 2.5m long and 20cm in diameter, packed with 45kg of ammonal. It was a race, now, to see which side could complete its deadly preparations soonest.

All sounds from the German workings had stopped (a sure sign that firing was imminent) by the time Westacott linked the exploder to the first torpedo's detonator, shouted a warning and pressed down on the handle. He hardly paused to examine the results, but instead rushed to where the wires leading to the second charge terminated, and blew that too, bringing down tonnes of debris.

Perhaps a less experienced man would have let it go at that, but Westacott knew how fast professional miners could dig their way through such a fall; he directed his men to tunnel into it, and break through into the German system, while he, his sergeant and three volunteers prepared to explore it, arming themselves with pistols and grenades, struggling into Proto apparatus and removing boots to cut down the noise of their approach. Twenty minutes later, the tunnellers broke through, and Westacott, Sergeant Brown and the three sappers crawled through into the unknown.

The German tunnel they found was better built than their own; more closely boarded, and with greater headroom. But it lacked any sign that German miners had been caught in the explosion. After a cursory inspection, the little party shuffled off down the long, dark gallery, feeling their way and stopping often to listen. After thirty-six metres, they came to a 'Y' junction and were faced with the choice of continuing on down its stem, and into the main German tunnel system, or turning right to explore the other branch, and gauge the damage they had done there. Westacott silently pointed to the right.

Another thirty-six metres of silent shuffling brought them to the far side of the heap of debris caused by the first torpedo, and here they were greeted by the dismembered remains of two German sappers and by the unmistakable sounds of their own men digging through to them. Safe, for the moment, Westacott settled down to wait, the bare bones of a plan falling into place in his mind. When the other members of his party broke through, he sent some back to pick up more mobile charges from the ready store near the tunnel's mouth while he led the rest into the darkness of the German workings. It was scarcely more than thirty minutes since the second explosion had ripped through the tunnels, and already the party of Germans he expected to come and investigate were late.

The Canadians had hardly reached the junction with the other arm of the tunnel when they saw, down the main gallery, the faint glimmer of a torch. Westacott whispered a command, and with just moments to spare the party was safely hidden in the darkness of the tunnel by which they had entered the workings, while the Germans – a party of seven, with an NCO at its head and a junior officer bringing up the rear – turned the other way, intent on determining the fate of their late colleagues but in fact heading for a confrontation with the Canadians who had been despatched to collect additional charges!

The sound of the Germans' jackboots covered what small noise the Canadians' stockinged feet made, as they fell in behind the unsuspecting men, and within moments, Westacott and his party were within striking distance – and none too soon, for at that moment the torches of the carrying party hove into view. The Germans stopped, no doubt supposing them to be their own men, and called out a greeting. The torches were extinguished, and after a moment's hesitation the German NCO at the front let fly with a pistol round. At that, Westacott switched on his electric torch, shone it into the officer's eyes from just a few feet away, and pulled the trigger of his .455-inch Webley revolver, shooting the other man in the stomach. Acutely aware that his back presented the best target for his own men, he flung himself to the tunnel's floor, and seconds later both heard and felt the roar of Sergeant Brown's revolver as he fired on the next man in the German party, and then stepped on his officer in his haste to get to the German NCO, kick the pistol out of his hand and administer a *coup de grâce*.

It was all over in moments, the Canadian carrying party's cries of 'For God's sake, shoot straight!' soon giving way to shouts of 'Kamerad' as the remaining Germans surrendered. Now, Westacott knew, there was no time to waste; a much stronger German party would soon be on its way to investigate the sounds of gunfire. The rest of the party made off, propelling its three prisoners, while he and Brown grabbed the two mobile demolition charges and headed off towards the main gallery, taking one charge forty-five metres past the junction and lighting its fuse, then doubling back to the junction and setting the other before heading back the way they had originally come. The charges went off, one after the other, just moments after they had reached the comparative safety of the roof fall, though it knocked them both off their feet anyway. Sometime later, Westacott returned to the German workings and set two more charges 'to make a proper job of it'.

The next two weeks passed relatively peacefully – save that Westacott himself took a week's special leave and was married in Kent – with the Canadians secure in the knowledge that the German counter-saps near their own mines were safely blocked.

By mid-September, that had all changed; now, the German workings were very close indeed to the Canadian galleries, and there were regular skirmishes once again as the two sides broke into each other's tunnels. During the night of 15 September, Westacott twice set off torpedo charges and captured short stretches of German tunnel, and was thus oblivious to the ferocious fighting which was going on above. Towards the end of the shift, he made his way to the surface, together with his sergeant, and was amazed to find the stretch of trench he had left twelve hours before in German hands! Both the Canadians and their enemies were totally overcome with surprise, but it was Westacott and the sergeant who reacted first, throwing themselves back down the mineshaft and hurling the satchel charges kept near the entrance for just such emergencies into the trench above, partially blocking it and the tunnel's mouth.

Frantically, Westacott sounded the alarm down the tunnel, and then, with the aid of a small party of men who rushed to his assistance, began to throw up a barrier of sandbags some ten metres in from the entrance, where the tunnel took its first right-angle bend. By the time the first Germans began to probe at the tunnel, the barricade was in place, and Westacott and his small party were able to hold them off with pistols and grenades; but soon enough the sheer weight of attackers carried some as far as the barricade itself, where they were cut down with knives, bayonets and digging tools. Meanwhile, Lieutenant Bill Robinson – the only other survivor of the Germans' late-May offensive, who was also down the tunnel that night – had begun to organize similar defensive barricades at all the other exits from the workings. And so the eighty men of the 2nd Canadian Tunnelling Company settled down to the most bizarre pitched battle of their long experience.

After four hours of fighting the combat soon settled down to a pattern: a new attack would be made about every twenty minutes, and would last for perhaps five, to be followed by a fifteen-minute lull – the Canadians were perilously close to exhaustion, and were scarcely able to breathe for the fumes which filled the tunnels. Here and there, as a torch shone out briefly, it would reveal a grotesque, horrifying scene: the bodies of dozens of dead and wounded Germans littered the tunnel before the sandbag barricades. And still they kept coming.

It was clear to Westacott that the barricade at the main entrance could not hold out much longer, and he now began to make plans to withdraw his tired defenders into the interior of the system preparatory to blowing the pre-placed demolition charges and bring the entire tunnel down for many yards, blocking the gallery completely. All went according to plan, and by a little after midday there was no further danger from that corner.

There was from virtually every other direction, though, and by 16.00 hours it became plain that a further retreat was necessary. Robinson started work on two more sandbag barricades, one each side of the apex of another right-angled bend, while Westacott saw to the placing of the demolition charges. That work was completed, and Westacott and his working party were between the first and second barricades when part of the former was blown out, and a German storming party forced its way through. A grenade landed near Westacott, and he threw up his left arm to protect his face. When he recovered consciousness, he found he had been dragged to comparative safety, and that the attack had been beaten off, but that he had lost most of his arm.

Sometime in the early evening, after almost twelve hours of fierce fighting, the Germans decided to withdraw, confident that they had the Canadians – whose numbers were now down to about twenty men capable of fighting on, most of whom were wounded to some degree, out of an original complement of eighty – penned in, and content just to contain them there until they could endure the conditions no longer. Had they known just how slim were the defenders' resources, doubtless they would have finished them off. In fact, they themselves soon came under attack, and the lost Canadian trench was recovered by British troops at first light the next day, and the gallant Canadians relieved.

In all, perhaps 50,000 allied troops – and at least as many German – were employed in tunnelling and mining operations on the Western Front, and very often were able to operate without giving away their presence to the enemy. In those circumstances, the explosion which resulted when the mine was detonated was all the more shocking for being completely unsignalled, and the raid or assault which invariably followed often had near-catastrophic results for the defenders (though only if the timing was accurate; the Hawthorne mine was blown ten minutes early, at 07.20, at the start of the Somme attack on 1 July, and proved a death-trap for the infantry sent to consolidate its effects).

Dunn gives us a clear picture of the sort of effect the exploding of an unsuspected mine had on the defending infantry: on 21 June 1916, the 2nd Battalion RWF, fresh from a rest period out of the line, arrived in the trenches at Givenchy to take over from the 4th

Battalion, the Suffolk Regiment. Captain Blair, in command of the Fusiliers' B Company, was assured by his opposite number that all the German mines in the area – which was notorious for the practice – had been 'blown to bits; not a sound has been heard since we blew the camouflets [counter-charges placed by one's own miners alongside the enemy's workings] yesterday.'[7] The reassurance – a repeat of one similar given by the officer commanding mining operations the previous evening, at a battalion officers' dinner – proved to be a hopeless misjudgement . . .

Just after evening stand-to, an immense explosion ripped through the battalion's position. Blair managed to half-free himself, but no more. In the company of Lance-Corporal Morris, who had suffered two broken legs, and the dead sentry, Bayliss, he passed the whole of the next day protected from German machine-gun fire by a slim parapet, a few inches high, and was eventually rescued the following night, to discover that he had been caught in the blast from the biggest mine the Germans ever exploded. 'Even on the broken ground of the Givenchy sector,' Dunn says, 'the new crater dominated the landscape. It was estimated to be some 120 yards long by 70 or 80 yards broad, its walls were upwards of 30 feet high. Figures vary according as internal or external estimates are given [*sic*]; the actual cup was some 40 yards across and about sixty yards from front to rear.'[8]

The mine was actually insignificant when compared to some; the remains of one of the biggest, the Lochnagar mine, near La Boiselle, is still visible today, the land in question having been purchased by a British businessman, Richard Dunning, to save it from becoming the site of just another *lotissement* (a small housing estate). Even after three-quarters of a century of erosion, its crater was still eighty metres wide and twenty-two metres deep.

Sergeant Roderick, of No 8 Platoon (B Company), was about his business of settling the men down for the night:

I hallooed to the CSM's batman as I passed (Pattison's dugout), and he called back 'Come down', but I passed on to my platoon only a few yards away. As I reached the first man there was a rumble, a tremble, an explosion. Up flew dust and soil and everything behind me, and then it fell about us; the CSM's dugout was filled in, and young Roberts (the batman) was never seen again. We were dazed a bit by that, but it was nothing to the

24

bombardment of the trench which followed. Jack Johnsons [a German medium artillery shell, which the French called a marmite – cooking pot; they gave off thick black smoke and were thus named after the heavyweight World Boxing Champion of 1908, the first black man to win the title] and all were dropping like manna. There were only eleven of us, in three bays of the Company's trench, not buried by the explosion ... When Gerry's [*sic*] guns lifted we could see his men coming over in three lines; there was enough light, especially when flares went up on someone's front to our left or right. They made a lot of noise talking and the white armlets they were wearing showed them up. We opened up with all our rifles and Lewis guns. We could see them being knocked over and carried away by their pals. They got confused and hesitated and made to come on in groups by the side of the crater. We fired whenever we could see anyone. One man came right round behind us; he was spotted by his armlet and was shot before he could heave a bomb ... Some of our lads climbed out at once and stripped him of his revolver and souvenirs [the impulse to loot was still demonstrably strong!], though I told them they were damned fools. While this sort of thing was going on Mr Banks [Roderick's Platoon Commander; a young man whose first time 'up the line' this was] suddenly ran across our front. I shouted to him, but he went for three Germans on the crater slope. One of them lifted his hand and struck Mr Banks before I could fire. Mr Banks' last effort was to jump into the trench. I ran up and finished with my bayonet the Gerry I had wounded. The other two ran away, letting go one of our men they had seized. When I got back to Mr Banks, he was dead, stabbed in the abdomen.

After a time I thought of getting in touch with someone on the right, so I left Lance-Corporal Davies in charge. I ran at once into four Gerries in the crater. I fired among them and downed one with my trenching tool, then I tripped over my rifle and bayonet. How I escaped being stabbed is still a mystery to me. They were between me and my platoon, so I had to follow the crater backwards ... When I got back our Lewis gun had been smashed by a minnie [*Minenwerfer*: German trench-mortar]; two of the team were killed and there were two dead Gerries beside it ... We might have captured some Gerries who had got lost in the crater, but we had all got a good shaking up and let ourselves be put off going for them. If there had been an officer about it would have been all right, someone to give an order and take no backchat.[9]

Stanway, also a Platoon Commander, whom we last encountered at the disastrous 25 April raid, rallied both his fellow officers and such men as were capable of fighting, 'pushing them out to the right, forming a flank that kept moving forward' (and freeing Lieutenant Craig, who was 'found captured, sitting in a shell hole with a large German standing over him – he did not stand long' in the process). With the coming of first light, Stanway 'got some idea of the havoc. It was a pretty ghastly sight. Six more or less buried men were seen and dug out of the crater slope; one of them died on the way to the Aid Post. There were a good few dead raiders lying about. From one *Unteroffizier* a map was taken which showed the whole scheme of the operation.'[10] This junior officer's laxity was to have extremely serious repercussions for his erstwhile comrades, as we shall see.

Meanwhile, Conning, the company's bombing officer, had been sent round the crater to reconnoitre. He found that No 5 Platoon had had a similar experience to that of No 8 – the Lewis gun and two of the team were missing. At first, they were thought to have been carried of as trophy (Lewis guns were much sought after by the Germans, who even produced an operating manual for them) and prisoners; about midday, however, the lance-corporal of the gun and a German were discovered dead beside it, in long grass between the parapet and the wire. During the morning, the remnants of the battalion set about repairing the trenchworks as best they could, and preventing the enemy from turning the new crater to his advantage.

It was during this work that Blair, still buried, along with Morris, in the crater wall, was spotted. It was not possible to mount a rescue operation then (a Cameronian in a neighbouring position had already been killed by a sniper while trying to reach the body of one of three dead German officers lying on the side of the crater, whose wristwatch he coveted) and the two men had to endure the longest day of the year in the open. Dunn himself led the rescue party which went out as soon as it was dark. Freeing Blair was a relatively simple matter, but Morris was more firmly interred, tangled up in barbed wire, torn sandbags, angle irons and one of Bayliss's legs. It was so near daylight by the time he had been released that there was no time for the party to recover Bayliss's body. When the battalion was relieved, on 27 June, the total casualty count for the six days spent in the line was 113, of which 93 had been caused when the mine went up and created what became known officially as the Red Dragon crater, after the Royal Welch Fusiliers' regimental crest.

The incident led to a more formal infantry operation, planned at length during the rest period which followed. The map taken off the German *Unteroffizier* showed the location of every dugout and feature in a salient the British called The Warren, and after some initial suggestion that an attack should be mounted there to take and hold the area, it was decided that a punitive raid in strength would be more appropriate. The Royal Welch returned to the line at Givenchy on the evening of 3 July, and the planned raid was launched as darkness fell on the 5th. Dunn takes up the narrative:

The idea is that D and A Companies, 200 in all, assault from the west and north-west respectively; that A push through to the reserve line and hold it, while D mops up and helps the Engineers of 11th Field Company to destroy all the tactical details: B Company, in parties of twenty each [it was B Company which had suffered the most casualties during the mine explosion and subsequent German raid, we may recall; forty was now its total effective strength], attached to A and D is to carry explosives and bombs [i.e., hand grenades]: C Company holds our own line . . . The CO's order impresses 'on all ranks that success depends on keeping a clear head and working at concert pitch during the operation'.

At dusk the assaulting Companies were ready to move into position under the barrage. At 10.30 the bombardment opened with field guns and 4.5 howitzers, 6-inch howitzers fired on known machine-gun positions; Stokes mortars and the Battalion rifle grenades also took part. D and A left their assembly positions when the guns began. 'They went away like a pack of hounds', said the CO, who was up to see D Company start. Using the cover given by the craters, D got as far forward into the very broken and wire-entangled ground as was possible and waited. A Company moved into Nomansland [*sic*] in single file. The ground scout who preceded the Company Commander and the CSM was shot through the head when he had gone only ten yards [9m]. Without further casualty or incident, the Company went eastwards for about 600 yards over soggy, marshy ground, through long grasses and bits of barbed wire, across abandoned trenches, and lay down, in close column of platoons, fronting the steep north face of the German re-entrant, about 120 yards away, having Mackensen Trench on its left. A white tape had been paid out to guide individuals and parties returning. [An unnamed observer adds: 'The men were very happy and soon had their cigarettes going. They were cracking jokes and calling Fritz a few choice names while the guns carried on the good work.'] At 11.15, when a box-barrage [an artillery barrage which effectively created an impenetrable three-sided *cordon sanitaire* around the enemy position to be assaulted], much denser than for our earlier raids, was formed around the area to be raided, the Companies rushed forward, each platoon making for its allotted place.[11]

Fox, A Company's Sergeant-Major, takes up the story:

I began badly: when standing on the [German] parapet it gave way and I fell in a heap at the entrance to a dugout, unable to help myself, having a Verey [*sic*] pistol in one hand and a rifle and bayonet in the other. 'Are you hit, Sergeant-Major?' Higginson [the Company Commander] asked. 'Yes, sir, with my own rifle.' Next moment I was calling the Boches from the dugout. The reply was a bullet, so the Mills bombs were thrown in. When the dugout was searched, four dead were found in it. Later an ammonal charge was placed at the mouth of it, and so they were buried.

The next dugout also contained four men. They came out with their hands up. Their belts and bayonets having been taken off them, they were told to get on the top and sit in a shell hole. One of them, speaking good English, said, 'Not for you, or your officer.' He got a wallop on the chin, and was hoisted up by his pals. Soon after, a man of ours came along with a wound of the left arm. He was given a revolver and told to take the four prisoners back to the Company's trench, and hand them over to the CQMS [Company Quarter-Master Sergeant] who was waiting in a deep dugout to receive prisoners. Those four never got to him. The escort wrote from hospital to a pal in the Company that he thought they were going to slip him, so he shot them.[12]

Dunn again:

A dozen half-stupefied occupants of the front line had surrendered at once to D Company. In the support and reserve lines, a stout

28

resistance was met; the conduct of one German officer was spoken of with admiration. A Company did most of the fighting, D the bombing and destruction. When resistance was overcome, looting and destruction followed. After D Company's CQMS had dumped the rations he waited for news, got impatient, and joined the raiders. So did men of B Company after carrying the extra stores. The German dugouts excited envy by reason of their solidity and comfort. As many as could be found were wrecked with explosives, so were the mine shafts, saps, trenches, and a mortar too large to carry away. Everyone had personal souvenirs. Among the captured weapons was a machine-gun of British make. It was carried down by Pte Buckley.

The weapon of Buckley's choice was a billhook; one of the wounded prisoners appeared to have been chased by him. Little Heastey too carried a billhook; two other officers carried a cosh (which Dunn, in his general glossary, describes as a nailed club). 43 prisoners were sent back, some of them wounded, and 15 identity disks were taken off the dead. The first batch of 25 prisoners came down under the sole escort of Moody's servant [Moody was the D Company commander]. They were Saxons of the 241st Regiment, a well-set-up lot of men. Our bombardment caught them when rations and mail were being distributed. After the strafe had lasted two hours in all the recall signal was made.[13]

The German artillery began retaliating as soon as the British bombardment opened up, and caused considerable damage and not a few casualties. Dunn, as MO, knew better than most what the effects of both raid and retaliation had been: 'Our casualties were one officer and six other ranks dead, one officer and upwards of forty other ranks wounded [and one man, a corporal from A Company, was captured]. Most of the wounds were from shell and bomb splinters, and occurred in A Company, whereas bayonet wounds were commonest among the German prisoners wounded.'[14]

And there we leave the 2nd Battalion, the Royal Welch Fusiliers, licking their wounds and showing off the spoils of battle, for further south the front, from Gommecourt as far as Chaulnes, across the Somme, had erupted in the biggest and most costly battle the British were ever to fight in the whole of that war, lasting from the first day of July until 14 November. We can leave to others descriptions of the artillery barrage and the way the murderous machine-guns were organized, but we do need to look, in passing, at the tactics of battle during this phase of the Great War, as opposed to those of the raid and the patrol, if only because the ultimate phase was still almost exclusively hand-to-hand fighting.

The Evolution of Set-Piece Battle Tactics to 1916

At Mons, during the Battle of the Frontiers in the war's first weeks, German infantry attempted to win the firefight from as far away as 600 metres, and were beaten by superior British marksmanship and fire discipline; elsewhere, French infantry was sent into the assault with little preparation and no protection and expected to carry the day by bravery alone. Quite clearly, neither of those methods stood any chance at all of success, and soon gave way to static defensive positions and the tactics which went with them. As the stalemate of trench warfare developed, there were various attempts made to break its grip by means of frontal assault – at Neuve Chapelle, Festubert, Aubers Ridge and Loos, particularly – but none suc-ceeded, though there were moments when it appeared that the new fighting methods developed over the winter of 1914–15 might actually work. A comprehensive analysis of that development pro-cess lies outside our ambit here (and the interested reader is directed toward Paddy Griffiths's *Battle Tactics of the Western Front*, and Hubert C. Johnson's *Breakthrough*), particularly since the apparent breakthrough never came.

At bottom, one was always presented with a stark dictum: a direct assault on a strongly entrenched position defended by well-sited and well-served automatic weapons was almost always (and the caveat is only included because, in war as in peace, one never really knows . . .) doomed to fail. Add copious quantities of barbed wire to the equation – and the German lines, in particular, were very well supplied from the beginning – and the task took on a particularly ominous quality. But there was no alternative . . .

Such tactics as there were came to rely on artillery barrage to cut the wire in front of the objective, destroy strong points and machine-gun nests and kill or demoralize the enemy to the point where he was incapable of resisting the assaulting infantry, who advanced across the intervening ground from their own positions in waves, a hundred metres apart and with five metres between each man, to take the objective at bayonet-point (the wave system was by no means ubiquitous at the Somme; some battalions advanced in something more resembling columns. In any event, as soon as they came under effective fire, the advancing men tended to group together and advance in spurts, from one shell hole to the next). To have assumed in mid-1916, during the build-up to the Battle of the Somme, that the artillery was capable of this massive task was clearly ludicrous; the types of ammunition available were not capable of delivering the required results (and production was deficient anyway; in March 1916, British munitions factories produced something of the order of 1.8 million shells of all types, and that was roughly equal to the entire production for 1915; the preparatory bombardment on the Somme consumed that number in less than seven days) and the guns were not available in sufficient numbers. The wonder of it is that it *did* work, sometimes, when all the factors combined and a huge helping of luck was stirred in; assaulting infantry did sometimes get into the enemy's trench system and even held portions of it. But the cost in human lives was astronomical.

The Battle of the Somme started at 07.30 on 1 July 1916. That is a fiction, of course, but a convenient one: for the planners at GHQ, back at Montreuil, it had started way back in the spring; for the artillery it started on 24 June with the opening of the preparatory bombardment; for much of the infantry it started soon after that, when they began to raid the enemy positions for intelligence purposes (by our earlier definition, the raids were actually reconnaissance patrols), both to assess the damage the artillery was doing and to gain an insight into enemy morale and preparedness. So as not to draw attention to the Somme sector, raiding activity was stepped up all along the front; some were very expensive affairs – in the Loos sector, two battalions lost a total of 15 officers and 385 men in a raid and the retaliatory barrage which inevitably followed. None the less, we will fix on 07.30 as our starting point, for it was then that the artillery fire ceased and the infantry battalions of the British Fourth Army, some 200,000 strong, 'went over the top' and walked straight into a living hell of machine-gun and rifle fire, to which was soon added artillery, switched from counter-bombarding the British trench system as it became clear to the German observers that the long-awaited attack was finally under way.

Some were simply cut down *en masse* when they were forced to bunch up together to pass through narrow gaps in what was otherwise quite intact wire, but it was almost as bad for those battalions caught in the open, in country which is predominantly rolling chalk down-land. Here, the front lines were far apart – over 600 metres, in some cases – and no man's land turned into a huge killing ground, as the German artillery added its weight to that of the machine-guns. At the pace they were constrained to make, it was to take almost half an hour for some detachments of heavily laden men to cross the intervening space; it is not surprising that very few of them made it. We can take as an example, perhaps, the 103rd (Tyneside Irish) Brigade, which went into the attack in completely open, almost featureless countryside around the village of La Boiselle, up the oddly named Sausage and Mash Valleys (the British awarded the strangest names to features of the terrain), parallel to the main road which runs straight as a die from Albert to Bapaume, towards the south of the British sector (and with the detonation of the Lochnagar mine as its salient feature, and the signal for the men to begin their advance). Intent on taking the German defences by storm, the brigade had almost a mile of open country to cover from its start line on the more poetically named Tara-Usna Ridge before it even reached the *British* front line, and was prey to German machine-guns every step of the way. Its four battalions lost 75 officers and 2,064 men – 71 per cent of their effective strength – many of them before they even reached no man's land. (Almost incredibly, a remnant of the 4th Battalion, on the very right, reduced to less than half a company's strength, managed to penetrate 650 metres into the German defences, and then continued towards the village of Contalmaison, almost a further 1,600 metres away, which was their actual objective, unde-terred by the fact that they were hardly more than a reinforced platoon in size, rather than a brigade, and effectively leaderless. But that is another story . . .)

The 34th Division's sector around La Boiselle was one of the worst in this respect, due to its very width, but the story was much the same right up and down the line. Those troops which did reach the German front line found widespread devastation, just as they had been led to believe would be the case (phrases like 'There won't be a German left in their line; our artillery's blown them all to Hell' were commonly to be found on every senior officer's lips), but it was illusory: the German defensive plan included provision of bomb-proof bunkers, up to ten metres underground, which kept the defenders safe throughout the preparatory bombardment. Paradoxically, though they were now in much greater strength than was usual during a raid, the British infantrymen were actually rather less able to make a good account of themselves – all had rifles and bayonets, not to mention heavy packs, and were thus much less agile than previously. Very few enjoyed anything like surprise, either, and many were debilitated by their passage across no man's land. Heavily laden though they were, their loads were not made up exclusively of offensive stores, and they soon began to run out, particularly of hand grenades. And the reinforcements they were expecting, which included parties carrying supplies, were to have an even harder time of it in no man's land now that the artillery fire falling there had been intensified.

All in all, the Germans were in much better shape than the British. The following account comes from Rifleman Aubrey Smith, serving with the 1st London Rifles in the 56th (London) Division just outside the northern limit of the battle zone proper, before Gommecourt. The 56th Division (a unit of the British 3rd Army) was involved in what had been termed a diversionary attack, though the men involved can have made little of the distinction:

> The ground to be covered was from two to three hundred yards across [it was actually rather more in most places. The German line followed the borders of Gommecourt Wood, the British the almost imperceptible ridge-line to the west; the two were not parallel], and the German machine-guns had a good field of fire before our boys reached their first trench. Many men fell from machine-gun bullets and others from shells, which the Germans were sprinkling about no man's land, but the waves reached the front line where there was bomb-fighting with the garrison. The front trench seemed to be smashed to atoms so the men passed on to the second objective, then to the third line, where Gommecourt was almost within their reach. Here they had to wait for reinforcements, reduced in numbers but quite full of fight. Time went on, however, and no fresh waves appeared, while on the other hand, aggressive parties of Germans were bombing their way up communications trenches from behind them. They had been lurking in huge dugouts under the ground which had survived the bombardment and were now disgorging bombers by the hundred. The first and second lines had not been 'cleaned

out' by the captors who, it should be said, had never been led to anticipate such huge underground caverns, and therefore took less notice of their unpretentious openings than they otherwise would have done. The German bombers swarmed all around the advanced elements, who soon began to realize that they were surrounded and trapped as the hours went by. There was nothing to do but use their bombs sparingly and wait for reinforcements.[15]

In fact, the 56th Division was to suffer from two factors: the inability of its neighbour to the north, the 46th, to make any headway in its sector, as it had encountered obstacles which effectively stopped its progress until all its smoke cover had been blown away, and a lack of tactical intelligence. The two divisions were tasked to attack the north-west and south-west margins of the wood and hook round to join up behind it, but the Germans had in fact fortified the point between their two axes of attack. Smith continued:

German machine-gunners and bombers, too, satisfied that the attack was definitely strangled north of the wood, came to the assistance of the enemy in our quarter. The bombardment from our side had now lifted onto Gommecourt [the artillery fire-plan was strictly time-dependent, and there was no way for the infantry to communicate its changing needs to the guns], consequently the Germans mounted their machine-guns on the parapets [of their strong point at the wood's apex] with a clear vision and perfect targets, while the wall of [their] shells, cutting off the attackers from further support, continued hour after hour. Again and again men started off carrying ammunition and bombs – the latter being a vital means of self-preservation in the captured trenches – but they were all annihilated.

The small bodies of our infantry who were holding on, having run out of bombs, used German grenades for a time until these were exhausted. Presently, the more advanced elements had to abandon their wounded and make a bolt to the rear. But the Germans were becoming more daring, knowing that our men were trapped, and they mounted their machine-guns in the open while their bombers, drawing on inexhaustible supplies, made it impossible for us to remain in their trenches. The run for life had to be attempted ... Through the intense artillery barrage and under heavy machine-gun fire men dropped right and left, until the ground between the trenches was an awful scene of carnage. The attack was over, and the best part of several battalions lay before our lines in this sector alone.[16]

With Bomb, Boot and Bayonet . . .

For close-quarter fighting, tactical doctrine had always set great store by cold steel. In the case of the British Army during the First World War, this meant a 45cm-long sword bayonet fixed to the Short Magazine Lee-Enfield (SMLE) rifle. The bayonet was as much a symbol – and an article of faith – as it was a statement of (very raw) fact, and the British saw to it that its virtues were inculcated into every fighting man. In a very real sense, they were embodied in the man responsible for that inculcation: Major (later Colonel) Ronald B. Campbell, DSO, Assistant Inspector of Physical and Bayonet Training, of whom Paddy Griffiths says: 'Few individuals on the Western Front managed to evoke such universal hostility among the literary elite.' One member of that elite, Siegfried Sassoon, said of Campbell's lecture:

He spoke with homicidal eloquence, keeping the game alive with genial and well-judged jokes. Man, it seemed, had been created to jab the life out of the Germans. To hear the Major talk one might have thought that he did it himself every day before breakfast [Campbell never saw action on the Western Front at all; an instructor of his calibre would have been far too valuable to have been wasted in the front line]. His final words were: 'Remember that every Boche you fellows kill is a point scored for our side; every Boche you kill brings victory one minute nearer. Kill them! Kill them!'

Clearly, elements of Newbolt's injunction to 'Play up! Play up and play the game!' were evident even at that late date. Bloodthirstiness was one of Campbell's consistencies. 'Your aim is to be bloodthirsty', he was fond of telling his audiences, 'and for ever think how to kill the enemy.'

Campbell and his entourage toured the units of the British Expeditionary Force (BEF) like a theatrical troupe, giving a highly polished, well-rehearsed lecture on the use of the bayonet and the spirit of the bayonet, employing such special effects as the technology of the day permitted, and finishing off by leading students through an assault course peopled with straw-stuffed dummies, each of which was to be stuck repeatedly, to the accompaniment of screams of hatred and the pulling of an appropriately warlike 'killing face'. Basil Liddell Hart apparently regarded the whole production as 'comic relief', but many found it inspirational, in one way or another, and it made a 'deep positive impact upon most of the lumpen rank and file at whom it was really directed'.

To revert to the philosophical for a moment: one is rather hard put to imagine a love (or even a liking) for the bayonet to be something capable of inculcation. Any problems the faint-hearted were likely to have had would have stemmed, after all, from the very personal nature of the weapon – though the same thing might be said of all forms of close-quarter combat. It seems far more likely that a very healthy respect indeed would result from even a cursory glance at the efficiency of the bayonet as an offensive weapon at close quarters (this was in the days before the development of the sub-machine-gun/machine carbine, one might add), and that that alone would be quite enough to secure its adoption. Motivation apart, there was (and still is) very little to bayonet training, beyond the simple rhythm of in, out, on guard; it is not, after all, a weapon which demands – or is even capable of demonstrating – much in the way of finesse. By the time of the Second World War, when the massed assault with cold steel had rather fallen out of favour, bayonet drill was covered in just one paragraph of the seminal *All-In Fighting* by Captain W.E. Fairbairn (though the section in question was actually contributed by Captain P.N. Walbridge):

The bayonet will be used in close hand-to-hand fighting where you have no time to reload, or more probably when your magazine is empty. Otherwise you would shoot from the hip or shoulder. Except when in close formation among comrades, *keep the bayonet point low* [original italics]. In this position there is less chance of your thrust being parried and you are able to deliver a point in any direction. To make a point, lunge forward on either foot and drive the point of the bayonet into the pit of your opponent's stomach. Most of the upper part of the body will be covered by equipment. To withdraw, take a short pace to the rear as you wrench out the bayonet. You are then in a good position to deliver a second point, should this become necessary. If you are close to your opponent and unable to deliver a point, smash him on the side of his head with the butt, and follow up with the bayonet.[17]

Later, and at some length, we shall have cause to return to Fairbairn's work (his speciality was unarmed combat, but he was also the designer, with fellow Shanghai police officer E.A. Sykes, of the Fairbairn-Sykes Fighting Knife, the so-called 'commando dagger', as well as the heavier hybrid 'smatchet').

Not all close-quarter tacticians were as dedicated as Ronald Campbell to the pristine virtues of cold steel; some preferred the less personal (and, they argued, more versatile) bomb, and they were represented in sufficient numbers for a 'cult of the bomb' to have developed within the BEF by mid-1916 (not that the Old Contemptibles of 1914 had anything against bombs should any have come, as it were, to hand). It was the nature of the war, with its static defensive positions, which turned the hand-thrown bomb into a widely popular weapon – it gave the infantryman a means of indirect attack at short range; a bomb lobbed – or even rolled – into a trench or dugout was capable of great destruction in the relatively small enclosed space while not exposing the bomber to immediate retaliation. The combination of guns to keep the enemy's head down and bombs to destroy him while he was under cover was a potent one indeed. And if the bomb's usefulness made it popular, it was its popularity which made it into an effective weapon, as the munitions experts turned their professional attention to improving on what was often a soldier's expedient.

The earliest grenades were simply sticks of high explosive with a length of inflammable fuse attached, and the procedure for employing them was equally straightforward: a variant of the firework's instructions to 'light the blue touchpaper, and stand well clear'. Not surprisingly, they were less than universally effective in wet conditions, and at the Battle of Loos, for example, many thousands failed to detonate. Variants of this simple device were wrapped with heavy gauge wire or had large nails bound to them. The problems of actually setting the bomb off continued to plague designers until they perfected the simple time fuse in 1916 and, rather later, the 'allways' percussion fuse (so called because it was designed to function on impact no matter what the attitude of the grenade when it struck).

The earliest grenades were distinctly dangerous to the bomber, as well as to the bombed. Robert Graves was witness to an incident

in the Harfleur 'Bull Ring' (as the training camps set up in France were known) when an instructor tapped a percussion fused grenade on a table for effect, and wounded fourteen men in the resulting explosion. Sometimes the effects of even a simple accident could be catastrophic: young Billy McFadzean was a bomber in the 14th Royal Ulster Rifles. Just before Zero Hour on the Somme, he and his fellows were huddled in a narrow assembly trench in Thiepval Wood (properly known as the Bois d'Authuille) making their final preparations. Suddenly a box of grenades fell to the floor, and the fall knocked the pins out of two of the bombs. Most of the men froze, horror-stricken, but McFadzean pushed forward and threw himself down upon the grenades. He was awarded the Victoria Cross posthumously for an act of selfless heroism which a small improvement in technical design would have rendered unnecessary.

If there was little finesse to employing the bayonet, it was a true skill compared to the work of the bomber, whose only consideration was to make sure that he and his colleagues were out of the blast pattern (best achieved by making the bomb's lethal radius quite small). In practice, certainly in clearing trenches or bunkers, personal safety was seldom a consideration, the blast and fragments being naturally contained within the confines of the objective, and many bombers would have preferred more powerful grenades. Rifleman Cantlon of the King's Royal Rifle Corps was one of them:

> [We had] no training, as such, except that we were supposed to chuck bombs at these flags that were supposed to be dugouts. Well, when we got to the real thing and we were supposed to throw them down real dugouts full of Germans when we got into the trench, the first thing was that the bombs weren't nearly powerful enough to do much damage.[18]

Enterprising bombers converted 11-lb Stokes mortar bombs and used those in place of grenades; others, particularly those tasked with 'cleaning out' dugouts, many of which were built with a 180-degree turn in their access stairs, to act as bomb traps, tried wiring a bomb to a jerrycan of petrol, but the resulting device, though certainly effective, was cumbersome in the extreme. Incendiary grenades, filled with 'Thermit' – a mixture of powdered aluminium and iron oxide – or phosphorus were also available. (Flame-throwers were also to be found at the front by the time of the Battle of the

Somme, but they can hardly be classified as infantry weapons: each one weighed over two tonnes. At a little under a hundred metres, their effective range was too short to be effective, and they were impossible to manoeuvre over broken ground; German experiments in both man-portable flame-throwers, for use in the assault role, and static versions, to be sited in defensive positions, were more fruitful. By one more curious twist of fate, the man who led the first effective *flammenwerfer* unit was the ex-chief of the Leipzig fire brigade, Herman Reddeman.)

A bomber with a good arm was effective at twenty metres or more; for longer ranges, a simple launching device powered by an uprated blank cartridge was available for the SMLE rifle, firing a standard Mills Number 36 grenade modified by the addition of a simple baseplate (the actuation method of the Mills was well adapted to this method of discharging it: the lever which held the striker was retained in a cup or ring attached to the rifle's muzzle, with the safety pin removed; when the grenade left the cup the lever was thrown clear in the normal way, the grenade was thus primed, and exploded five seconds later). The range of the early launcher was between 50 and 200 metres.

By early 1917, each British rifle platoon had a section equipped with grenade launchers and another detailed as hand bombers. In the attack, the rifle grenades and the Lewis gun (see page 43) gave fire support while the hand bombers and the riflemen made up the assault element. The combination of rifle grenades, light automatic weapons and rifle fire has gone on proving effective right up to the present day, though the weapons themselves have changed out of all recognition, and the principle of organic fire support in units of all sizes, from the section/fire team up to the battalion and beyond, forms the basis of modern infantry tactics.

New Weapons and Tactics

If we have concentrated on the British Army (and that included colonial units under what had become a very wide banner indeed) thus far, it is simply because English-language sources are easier for an anglophone to deal with; in fact, the Austrian, French, German, Italian, Russian and Turkish experience (to name but the major combatant nations at a time when the USA had still to enter the war) was hardly different – they were meeting the same problems in much the same way with very similar weapons and tactics. On the Western Front, the only important difference was in the nature of (some) French and German defensive emplacements, which were most definitely built to last, while the British and (most) French had clearly made the decision that they were not going to be where they were for very long, and would see out their short stay in temporary structures.

The German policy of massive expenditure on defensive construction became clear during 1916, first of all during the summer, on the Somme, and later, in November, when they made a studied withdrawal to the Siegfried/Hindenburg Line, giving up all the territory the British had hoped to take during the summer/autumn offensive in return for secure, comfortable quarters and virtually unassailable positions protected by fields of barbed-wire entanglements hundreds of metres deep. By then, the British had evolved a better way of cutting the wire by means of high-explosive shells, fused to burst at the precise moment of impact, whose blast pattern was horizontal, where previously they had depended on time-fused shrapnel shells bursting before they hit the ground; as we have seen from the accounts of the Battle of the Somme, that was very much a hit-and-miss affair. The new Mark 106 fuse was a significant improvement, but it still could not guarantee to clear a path for the advancing infantry, particularly through wire in the sort of depth found before the Siegfried Line, and it did nothing to nullify the power of the machine-gun. What was needed was a way to combat both the local elements of the enemy's defensive screen – wire and machine-guns – and allow the infantry a fighting chance of getting to a point where they could assault his strong points, bunkers and trenches with bullet, bomb and bayonet.

The tank (the name came about as part of an exercise in deception, designed to keep the new armoured fighting vehicles from the enemy's attention; the new constructions were said to be mobile water tanks, bound for the deserts of Mesopotamia, and the name stuck, in English at any rate) was a comparatively long time coming, considering its relatively low level of technological sophistication: all the elements (the 'caterpillar' track was the last of them to appear) were in place as early as 1910, but it was 1916 at the Somme before the new weapons first appeared on the battlefield, and then they were introduced prematurely. They were by no means reliable, even by the standards of motor vehicles of the day. Accordingly, they were not an outstanding success at first, and during their next major deployments, at Arras, Messines and the third battle of Ypres, the following year, they were insensitively employed and proved to be next to useless in a battlefield which had been 'prepared' by extensive artillery bombardment and then turned into impenetrable waist-deep mud by the worst weather for decades. It was not until November 1917, at Cambrai, that the Tank Corps got a chance to operate as it wished, on a battlefield untouched by artillery barrage, and it scored a huge success – though one which, sadly, the much depleted infantry were totally unable to follow up. The experience at Cambrai proved two things: that tanks could overcome even the best-prepared defences, and that they could no more win battles on their own than could the aeroplane.

They also gave the infantry an unwelcome new task – tank killing. At first, without adequate weapons, the best that individual German infantrymen could do – almost incredibly – was hang on to the guns (six-pounders and Vickers, Hotchkiss or Lewis machine-guns) to prevent them by brute force from being trained on a new target. Individual hand grenades were completely ineffective unless they could be delivered through firing slits; soon enough, however, German grenadiers learned that a sack of bombs was likely enough to disable a tank, particularly if it could be placed in or under a tread, whereupon the tank could be destroyed by gunfire from field artillery. An officer from the 46th Infantry Regiment described his men's reaction to the appearance of tanks on the battlefield, and how they fought them:

> It is found that they [the tanks] are able to conquer ground but not hold it. In the narrow streets and alleyways [of a town or village] they have no free field for their fire, and their movements are hemmed in on all sides. The terror they have spread among us disappears. We get to know their weak spots. A ferocious passion for hunting them down is growing ... As individual hand grenades, thrown on top of the tanks or against their sides are ineffective, we tie several grenades together and make them explode beneath the tanks.[19]

Lieutenant Spremberger of the 52nd Reserve Infantry Regiment was involved in the defence of Fontaine, on the road from Bapaume to Cambrai, one of the last objectives the tanks overran during the November battle. He tells a similar tale:

At first we tried to throw hand grenades under the tracks of the tanks, but their explosive power was too weak [to have any effect]. So I ordered empty sandbags to be brought up. In each bag we placed four grenades ... Musketeers Buttenberg and Schroeder, both members of our assault group, ran towards one of the firing monsters and put two of the bags under the track. A violent explosion followed, and the left-hand track was blown completely off its wheels.[20]

It was to be another twenty-five years before the infantryman had a more effective weapon with which to fight tanks – rocket-powered projectiles such as the Panzerfaust and Bazooka – but even after their development, anti-tank grenades still proved workmanlike and – perhaps 'popular' is not quite the word – widely used. True to the old dictum that like best combats like, guns such as the superb German 88mm of the Second World War, which were used both to equip tanks and as towed anti-tank guns, and the British six-pounder were to prove the most efficient means of all of destroying armoured vehicles until the advent of guided weapons in the 1960s. Small calibre anti-tank rifles, such as the 13mm calibre 'T-Gew' (Tank-Gewehr) introduced right at the end of the Great War, were also produced, but proved unpopular and of strictly limited value; the infantrymen delegated to fire them did not mourn their passing.

If the tank provided the assault infantryman with a means of breaching the wire and getting to the front line proper of the enemy's defensive positions, it still left him with the problem of suppressing enemy defensive fire at close range and clearing trenches and bunkers (another pair of euphemisms for killing the defenders). What was needed, in the opinion of many of the men who had become experts on this subject in a very short time indeed, was increased firepower, and that led to the introduction, first of all, of lightened versions of existing machine-guns and then the development of a new form of automatic weapon entirely, the so-called sub-machine-gun, machine pistol or machine carbine. We have no room here to examine the technical development of such weapons in detail, but instead propose to concentrate on the way they were employed, for they were destined eventually to change the nature of close-quarter combat considerably.

No one, on either of the opposing sides, would have suggested using machine-guns in an assault role during the early part of the Great War. The British and French, it is true, had finally recognized that the machine-gun was essentially an infantry weapon, and not part of the artillery, but they were still devoutly convinced – for the most part – that it was only to be used in defence, and from fixed positions. The nature of the machine-guns themselves seemed to confirm this: they were impossibly heavy, on their massive tripods, for rapid tactical movement, and required careful orientation if they were to operate successfully, particularly as indirect-fire weapons at the extremes of their range.

Such 'light' machine-guns as did exist, notably the Danish Madsen (which the British had rejected, and which the Germans captured in some numbers on the Eastern Front, and later used in action on the Western) and the French Hotchkiss Modèle '09 and Chauchat, were generally unreliable and – particularly in the case of the last-named – wildly unpopular. There was an exception though: the American Lewis gun, which was manufactured in large quantities in Britain from 1914 onwards, even if the machine-gun purists dismissed it as a mere automatic rifle. The Lewis gun was light enough – just – for one man to operate it with no more of a mounting than a simple bipod (indeed, there are instances of the gun being fired from the shoulder, like a rifle), and fitted with a sling, it was quite feasible for a big, strong man to fire it from the hip, though it was 1918 before an operating manual for the gun got around to recommending that.

It was its portability which made the Lewis gun so popular with the German infantry, which utilized captured examples in such numbers that a German-language operating manual was soon produced, and the lesson was not lost on the *Gewehr-Prüfungskommission* (the body responsible for German small arms procurement). In late 1916, a lightened (but only by 4kg) version of the standard German MG08, known as the MG08/15, went into production at the Spandau armoury. The 'new' weapon was first issued generally in the Verdun sector in early 1917. Like the Lewis gun, it had a conventional butt-stock, a pistol grip and an integral bipod, but its ammunition supply was still in the form of a belt, albeit now contained in a drum, in place of the Lewis's more convenient pan magazine. In fact, there was really nothing new about it at all; it was simply a cut down Maxim gun as modified by German arms manufacturer DWM in the opening years of the century, and suffered from all the Maxim's shortcomings. There were twenty-one possible reasons for the gun stopping listed in the operating manual. It was adopted in place of the arguably more effective air-cooled Bergmann MG15nA or the Parabellum-MG, because the mass production tooling was already in place, and could be quickly and easily duplicated. In all, 130,000 examples of the new gun were to be produced before the war's end. They were to become associated (though for no particularly good reason; they were issued to all infantry battalions) with one particular arm of the German infantry: the *Stosstruppen*, or stormtroopers.

To break the deadlock of trench warfare, the Germans developed radical new units and tactics. To lead the way through the enemy trench lines they created the stormtroops. These were small units of intensively trained and heavily armed troops, backed by highly mobile artillery. The stormtrooper was specially equipped for his work on the front line. Steel helmets came as standard, although steel body armour was not adopted because its heavy weight had a detrimental effect on the stormtrooper's mobility. Wire cutters were carried to clear paths through barbed-wire entanglements, along with stick grenades with long wood handles which could be thrown to ranges of up to 40m. Personal weapons were selected to maximize firepower and minimize weight. The first sub-machine-gun, the 9mm Schmeisser MP18, was readily adopted by the stormtroopers, and most of them also carried a pistol for close-range work. These included the famous Pistole 08 Luger and extended-butt Artillery Mode Luger and Mauser C96.

The Germans realized that the old tactics of massed bombard-ments and bayonet charges were futile, so they dispensed with the heavy artillery barrages to achieve surprise. Small groups of storm-troopers were sent against weak spots in the enemy front. 7.62cm *Infanterie Geschutz* light field guns fitted with armoured shields were manhandled up to the front line to blast enemy machine-guns to pieces at short range. *Granatenwerfer* light grenade launchers, flame-throwers and MG08/15 light machine-guns joined in this pinpoint bombardment of the linchpins of the enemy's defensive line. The stormtroopers would use ground to hide their approach and attack from unexpected directions. Assault sections used fire and manoeuvre tactics to ensure that weapons were always trained on the enemy as stormtroopers moved in the open. Once the enemy's machine-guns were destroyed, it was then relatively easy to cross any barbed-wire entanglements. Stormtroopers would then move from trench to trench, throwing in hand grenades or bayoneting the hapless occupants. On occasions the stormtroopers used local-ized gas attacks to spread confusion in allied units.

Once a gap had been found, the stormtroopers would surge forward into the rear of the main enemy defensive position, spreading chaos and confusion. By attacking from the rear, the allied trenches and barbed-wire entanglements were almost useless. As long as the stormtroopers kept moving, the allies' field-telephone-based artillery fire control system could not coordinate effective barrages fast enough.

To carry out these tactics successfully, the Germans recruited specially trained squads of very fit and determined soldiers. Unlike the old regimented tactics, stormtroop tactics relied on initiative and improvisation to be successful. This meant NCOs were given unprecedented responsibility in battle. They had to be able to understand the attack plan and judge how the battle was developing to ensure unforeseen enemy resistance was quickly overcome.

By 1917, every German division had its own stormtroop battalion which was tasked with training every soldier in the division in stormtroop tactics. In the 1918 Spring Offensive, the stormtroop tactics worked well at first, driving all before them. However, the British were soon able to reorganize and build new defence lines before the Germans could bring up new supplies to exploit the success.

The German Army had been experimenting with specially trained and equipped assault troops since early 1915, and impressed with the success of the first *Sturmabteilung* (a name later to be revived by the Nazi Party, see page 57), created out of the 18th Pioneer Battalion, the German General HQ (OHL – *Oberste Heeresleitung*) ordered each of its fourteen armies to form a *Sturmbattalion* on 23 October 1916. These battalions were not the only specialized assault troops within the German ranks; since mid-1915, individual regiments – they had much greater autonomy than units of the British or French Armies – had been creating small units of their own (though they were known to operate even in company strength, and in the German Army that meant up to 280 men), known as *Sturmtruppen* (assault troops), *Jagdkommando* (hunting unit) and *Patrouillentruppen* (raiding troops). They were some of the first to receive MG08/15s and the even handier but never-quite-perfected Hugo Schmeisser-designed MP18/I machine pistol, first of a new breed of automatic weapons which would very soon take pride of place in close-quarter battle. They were also issued with improved grenade launchers and effective man portable flame-throwers (and also enjoyed better rations and living conditions than

45

the line infantry units; they were elite in every sense of the word). By the time of the battles of 1917 – Arras, Messines and 3rd Ypres/ Passchendaele – they were very proficient fighting units, and deserve most of the credit for the very successful counter-attack at Cambrai and the breaking of the Italians at Caporetto, where they were sent to stiffen the Austro-Hungarian Army. But it was in the break-through battles of spring 1918 that they really came into their own, by which time they were well used to their unenviable task:

We had the enemy under our fire [and] they gave way before us, so that the short stretches of ground were soon heaped with corpses. It was a nerve-scourging spot. We dashed over the still warm muscular bodies, displaying powerful knees below their kilts, or crept on over them. They were Highlanders, and the resistance we were meeting showed that we had no cowards in front of us.

After we had gained a few hundred metres in this fashion we were brought to a halt by the bombs and rifle-grenades that fell more and more thickly. The men began to give way.

It is just in trench fighting, the fiercest fighting of all, that such recoils are most frequent. The bravest push to the front, shooting and bomb-throwing. The rest follow on their heels automatically, in a herd. In hand-to-hand battle the fighters jump back and forward, and in avoiding the murderous bombs of the enemy they run back on those behind them. Only those in the forefront know what the situation is, while further back a wild panic breaks out in the crowded trench. Indeed, if [the enemy] seize their opportunity all is lost; and it is now for the officer to show that he is worth his salt . . .

I succeeded in getting together a handful of men, and with them I organized a nucleus of resistance behind a broad traverse. We exchanged missiles with an invisible opponent at a few metres' distance. It took some courage to hold your head up when they burst and whipped up the heaped soil of the traverse. A man close to me shot off cartridge after cartridge, looking perfectly wild and without a thought of cover, till he collapsed in streams of blood. A shot had smashed his forehead with a report like a breaking board . . . I seized his rifle and went on firing.

We tried several times to work our way forward by crawling flattened out over the bodies of the Highlanders, but we were driven back each time by machine-gun fire and rifle grenades.

Every casualty I saw was a fatal one. In this way, the forward part of our trench was gradually filled with dead; and in turn, we were constantly reinforced from the rear. Soon there was a machine-gun behind every traverse. I stood behind one of these lead squirts and shot till my forefinger was blackened by smoke. When the cooling water had evaporated, the tins were handed round and to the accompaniment of not very polite jokes, and by a very simple expedient, filled up again.

A cool evening breeze promised a sharp night. Wrapped in an English trench-coat I leaned against the side of the trench and talked to little Schultz, who had turned up with four heavy machine-guns just where there was most need of them. Men of all companies sat on the fire-steps. Their features were youthful and clear-cut beneath their steel helmets. Their leaders had fallen, and it was of their own impulse that they were here and in their right place. We set about putting ourselves in a state of defence for the night. I put my pistol and a dozen English bombs beside me, and felt ready for all comers, even though they were the most pig-headed of Scotsmen.

Then there came a new outbreak of bombing from the right, and on the left German light-signals went up ... Everyone seized his rifle and rushed forward along the trench. After a short encounter with bombs a body of Highlanders made hurriedly for the road. There was no holding us now. In spite of warning shouts, 'Look out! The machine-gun on the left is still firing!' we jumped out of the trench and in an instant reached the road, where there was a stampede of Highlanders. A long, thick wire entanglement cut off their retreat, so that like hunted game they had to run past us at fifty metres' distance. On our side there broke out a tumultuous hurrah which must have struck their ears like the trump of the last judgement, and a hurricane of rapid fire. Machine-guns mounted in haste made a massacre of it.[21]

By the time the Great War came to an end, on 11 November 1918, the armies on both sides had changed out of all recognition from those which had started the war, over four hard years earlier. They still used the same basic infantry equipment, but their tactics had undergone essential revisions, in particular the substitution of 'infiltration' tactics for the all-out frontal assault, seeking out weak points in the enemy's defensive line and concentrating the attack there, using artillery fire much more precisely, and not just blanketing an entire area and hoping for the best. It was just these tactics, plus the involvement of armour in vastly greater quantities, and the intro- duction of close air support, which were to dominate during the next and much enlarged global war which engulfed first Europe and then the whole world scarcely more than two decades later.

Chapter Two

A Violent *entr'acte*: Rearmament and Civil War in Europe

The repeating rifle and the heavy machine-gun dominated the infantry battlefield throughout the Great War, but by its end they had been joined by a range of new weapons designed to be effective – largely by virtue of being handier – at close quarters: the sub-machine-gun (also known as the machine pistol and machine carbine), the light machine-gun and the anti-personnel and anti-tank grenade, together with combat shotguns, flame-throwers and semi-automatic pistols.

The first sub-machine-gun, which came to be known as the MP18, appeared in 1917, the work of the German designer Hugo Schmeisser for arms manufacturer Theodor Bergmann. It employed relatively low-powered pistol calibre 9mm ammunition which permitted the use of a simple direct blowback/unlocked breech system of actuation. This system relies on the pressure of a large forward-acting coil spring to keep the breech-block pressed tight against the round inside the chamber; firing the round causes the now-empty cartridge casing to be propelled backwards, and when it overcomes the pressure of the spring, the result is to push the breech-block to the rear, allowing the casing to be ejected. Once the pressure on the spring falls, it returns to its original position, carrying with it a new round which it has stripped out of the magazine. It is uncomplicated, and therefore both cheap to manufacture and reliable even in quite adverse operating conditions, and Schmeisser's prototype set the standard for a new close-quarter weapon which put significant firepower, quite literally, into the individual infantry-man's hands for the first time. Neither it nor its successors was particularly accurate, but that was never an adverse factor.

Schmeisser's new gun was an idea whose time had come. Already,

49

the American, Colonel John Tagliaferro Thompson, who had begun work on a design for an automatic rifle to rival Browning's (see below), had switched his attention to a close-quarter weapon he nicknamed the Trench Broom, using the heavier .45-inch calibre ACP/M1911 round. This soon appeared as the eponymous Thompson Gun. Meanwhile in Italy, Abiel Revelli had produced a 9mm calibre blowback-action machine-gun for the Officina Villar Perosa some two years previously, though this, admittedly, was conceived as a vehicle-mounted light support weapon, and did not appear reworked into the form of a machine carbine until 1918. Only the MP18 saw action during the First World War, and indeed, it was to be fifteen years more before the development of the genre restarted in earnest with the revival of the German armaments industry following Hitler's ascent to power. In the intervening period, however, German manufacturers, including Bergmann, got round the ban imposed by the Treaty of Versailles and produced new models in small numbers in foreign factories, while Beretta was active in Italy and Lahti in Finland. Russian designers later drew heavily on the work of both the Finns and the Germans in producing a variety of designs culminating in the rugged and reliable PPSh41.

Essentially, the sub-machine-guns of the 1930s were all very similar in nature and differed chiefly in the way their magazines were mounted and functioned, and it was not until further cost and time constraints came to be applied to their manufacture during the Second World War that 'cheap and cheerful' designs such as the British Sten, the American M3 and the rather better-made German MP40 dispensed with wooden furniture and the luxury of machined, rather than moulded, metal parts, earning the sub-machine-gun an unenviable reputation for unreliability in the process.

The Lewis gun, invented in the USA and manufactured both there and in Europe in great numbers during the Great War, was that period's best-known – and best-loved – light machine-gun (LMG), and gave credibility to the entire genre of 'light' support weapons. Air-cooled and magazine fed, it could be operated satisfactorily by just two men. Admittedly, unlike the heavier machine-guns such as the Vickers, Browning and Maxim, it was incapable of true sustained fire – no LMG is – but it met the need for a fully automatic weapon – light and handy enough to go into the attack with the advancing infantry which the heavier guns simply could not. Together with its near-contemporary, the

Browning Automatic Rifle (BAR), it was still in service in the Second World War, while the BAR, though it was never the versatile weapon the Lewis was, soldiered on for a further decade.

Though a variety of weapons had been tried over the preceding decades, it was not until the interwar period that a European design for an LMG really worked. By then Germany's armed forces, for one, were turning to what we have since learned to call general-purpose machine-guns, light enough to go forward into the attack yet also robust enough for the sustained-fire role. The best of the interwar LMGs was the Czech ZB vz26, which, in improved form, was adopted by the British as the Bren gun and manufactured into the 1960s, though there were gas-operated, box-magazine weapons in most arms-makers' catalogues.

True sustained-fire weapons had been made possible by the development of the drawn brass cartridge, which also went on to make profound changes to the design of the pistol. The ubiquitous revolver now became much easier to load, and was soon joined by the so-called 'automatic' (actually, semi-automatic) pistol, designed by the Feederle brothers, and manufactured at the close of the nineteenth century by the German company Mauser-Werke, as the 7.65mm C96.

Over the first decade of the new century, many more self-loading pistols arrived on the market, the most famous of them being the Luger P08 and the Colt M1911, as well as the first offering from a man who did as much as any to revolutionize the design of automatic weapons, John Browning. British gunmakers Webley, better known for their heavy .455-inch calibre military revolvers, also produced a self-loading pistol, and so, before 1920, did virtually every other arms manufacturer then extant, in a variety of sizes and shapes, and in all sorts of calibres from .22-inch up to .455-inch. Very few had anything but a straightforward recoil action, though some, like the much sought-after Luger, used a version of it which became known as the Parabellum action – an upwards-breaking locking toggle not at all unlike the action of the original Maxim machine-gun. This was hardly surprising, since the pistol was developed by DWM, the Berlin-based company which first imported, and then itself manufactured, the heavy sustained-fire weapon.

Throughout the First World War the British, French and Russian armed services were issued with revolvers – Webleys, Lebels and Nagents, respectively – while the Americans, Germans and Italians were provided with automatics; Colts for the former, a mixture of Lugers, Mausers and often-unreliable Langenhans for the German Army and Navy, and Berettas for the latter. Among the smaller nations, Spain deserves a mention for the Astra, Star and Llama models, which all became available during the interwar period – the Star, like the Beretta Modello 23 and the Mauser C96, was even sold, albeit briefly, in a fully automatic version.

Arguments over the relative vices and virtues of the two distinctly different types of pistol, revolver and semi-automatic, have raged ever since the automatic was invented, though there is no denying that in close-quarter battle the greater ammunition capacity of most automatics – and particularly, perhaps, that of the Browning GP35, also known as the High Power, the bulky magazine of which holds no less than thirteen 9mm rounds – was a telling factor.

Shotguns – usually double-barrelled weapons with their barrels shortened to around 38cm, and with the butt lopped off to form a short pistol-grip – found some favour during the vicious trench fighting of the First World War, and variants were also to be found in the infantryman's armoury in the Second World War and later. They differed from their sporting counterparts not just in the surgery used to shorten them, but also in the loading of their ammunition: the No 4 cartridge, containing a dozen balls of about 7mm diameter, was – and still is – the preferred lethal loading; each cartridge has the same effect, at close range, as half a magazine from a sub-machine-gun. Some men preferred to reload their cartridges encapsulating the shot in candle wax, so that separation would also be delayed. More recently, a wider variety of cartridge loads has become available, from solid shot to so-called 'rubber' bullets, giving the shotgun a much wider appeal. These latter require the use of a conventional round-section barrel, but cartridges loaded with sub-calibre separate shot often have barrels with an ovoid section to ensure horizontal, rather than homogeneous circular, distribution patterns.

The breech-loading shotgun's main drawback was its very limited ammunition capacity – two shots before reloading, no matter how quickly the operation can be carried out, is seldom enough in any combat situation. They soon gave way to pump-action or semi-automatic guns with a five- or more round capacity, and, more recently still, to weapons with a detachable drum or box magazine in place of the more usual fixed tube under the barrel. In combat, the shotgun's main drawback is its very limited lethal range – forty to fifty metres is the absolute maximum. None the less, where this is less of a problem – in urban areas, and especially in house clearing, for example – they continue to play a useful role despite their relative bulk at a time when weapons are generally decreasing in size.

As we noted earlier, the anti-personnel grenades of the early days of the First World War were often expedient weapons, made up on the spot from a variety of objects which were relatively easy to come by on the battlefield. The simplest were fashioned from empty ration tins filled with recycled explosives from undetonated shells, fitted with the most straightforward time-delay mechanism possible – a length of slow-burning fuse which was lit with a match or by applying the lighted end of a cigarette; sometimes they were bound around with large nails, or had quantities of scrap iron embedded in them.

The grenades of industrial manufacture then available, such as the Hale's grenade, made in Britain from 1908 onwards in small quantities, tried to overcome the one main shortcoming of both the improvised bombs and those previously manufactured (which was that the fuse frequently failed to burn through; unsurprisingly, in wet conditions, many failed simply because their fuses were extinguished) by substituting a chemically ignited fuse for the length of slow-match, a process which was found to be far from straightforward. Percussion fuses were the most popular initially, the problems of ensuring that the grenade landed in such a way as to detonate the percussion charge being solved either by fitting a multitude of strikers or by stabilizing the grenade in flight by means of cloth streamers so that it tended to hit the ground the right way up. The other means of protecting the grenadier from the force of the explosion was to equip the bomb with a manually initiated time-delay fuse, and after some initial failure, this was the method most widely adopted during the First World War and through the interwar period. The grenadier either removed a safety pin and threw the grenade, a detonator-retaining lever held in place by the pressure of his hand being jettisoned when the grenade was released (the method adopted by the British for the Mills No 36 bomb), or he pulled a cord attached directly to the detonator before throwing the grenade (the method used in the German stielhandgranate). Both these types stayed in operation through the Second World War, though by that time they had been joined by more effectively fused percussion bombs.

When tanks first appeared on the battlefield in 1916, they were to all intents and purposes immune to single grenades, and the only real chance an infantryman stood of disabling one was to deliver multiple grenades right onto the vehicle's tracks, as we have seen. There was no successful anti-tank grenade as such produced during the First World War, but by 1939 a variety of more-or-less ingenious devices had been developed, most of which relied on an updated version of the Hale's grenade's cloth streamer 'tail' to land it so that the shaped charge incorporated into the grenade operated in the right direction. One exception was the British No 74 (ST) bomb, which was a spherical device with a short protruding handle, the whole of its curved surface being coated with a strong adhesive. As issued, the No 74 bomb had two hemispherical covers, which were removed before it was thrown. It was not a marked success.

More recently, the technology of grenades has improved considerably, and we shall deal with the new types in due course. One of the older types of non-high-explosive grenade, and one which was used extensively during the First World War, was the incendiary bomb. From very ancient times it was widely recognized that flame was a potent weapon, and as early as 500 BC, attempts had been made to project it. By 1916, there were a variety of quite effective flame-throwers in use on both sides, though only the Germans had yet perfected a man-portable version. The key lay in using a compressed gas to carry the inflammable liquid fuel: nitrogen, the most inert gas of all, proved to be the most satisfactory. Even large vehicle-mounted flame-throwers were severely limited in their endurance, however, and the early small back-pack versions were of very little use; it was only during the Second World War that more effective man-portable flame-throwers became available, and then they were used widely, particularly in neutralizing bunkers and strong points. The main difficulty in deploying them lay in finding men who were prepared to operate them – flame-thrower operators were seldom treated with any mercy at all, if caught, such was the horror the weapon produced in those subjected to it.

After the Great War, despite the attempts by the victorious Allies to deny Germany any easy route to rearmament by the draconian Treaty of Versailles, and their own sometimes naïve efforts to find a long-term solution to problems of keeping the peace through the establishment of the League of Nations, and by mutual agreements to disarm themselves, the two decades between the twentieth century's two world wars (or, as some historians would have it, during the extended break between its two parts) were a period of the most vigorous military expansion. Despite massive industrial reconstruction, the most profitable business to be in, all too often, was the manufacture of arms and munitions. And with the lessons of the Great War firmly digested, and new weapons of all sorts, from grenades to main battle tanks, ready to pour off the production lines at an unheard-of rate, the arms manufacturers were poised to reap their biggest harvest ever by the time the century was a third of the way through. To start with, though – and even while peace reigned – the future of ravaged Europe still depended on that most expendable of all commodities: the fighting man.

On Europe's eastern margins the Great War decayed into civil war without missing a beat, as Russia turned from fighting the Germans and instead began the lengthy process of tearing itself apart to create the Soviet Union. Apart from the broadest outlines, we really know very little of the way that war was conducted, but it does not require too lurid an imagination to picture the sort of horrific revenge the undisciplined White Russian armies of Kolchak, Denikin and Yudenich extorted when they advanced on the Reds' twin power bases, Petrograd and Moscow, in the spring of 1919 (with the oft-forgotten help and support of troops from Britain, Canada, France, Italy, Japan, Serbia and the USA, all of which sent armed men to fight alongside the counter-revolutionaries), or the way in which Leon Trotsky's Red Army behaved when it out-manoeuvred them, drove them back and annihilated them in turn the following year. The curtain the victorious Bolsheviks managed to draw around the detailed history of the events which followed is a very clear example of the old adage that the victors write the history, though news of some incidents, such as the uprising against Bolshevik methods at the naval base of Kronstadt in early 1921, which Trotsky crushed after fierce street fighting, did leak out.

Further west, four and a quarter years of static conflict had left behind it a poisoned wound which infected almost all of Europe, and reached out to the furthermost outposts of the British Empire and to the United States, too; only backward Spain and remote Scandinavia escaped entirely. It was not just individuals and the families of the huge numbers of killed, missing and maimed which suffered, but the basic infrastructures of all the combatant nations (save for the USA, which, having joined the war in time to be present for the *coup de grâce*, proved to be the only lasting winner, despite the interwar economic depression). In Germany, the biggest loser, the fabric of society soon proved to have rotted quite away, and was soon to be rewoven to a new and much stronger – though eventually no more durable – pattern under the hand of Adolf Hitler.

Foremost among the practical elements which contributed to Hitler's rise to power was his ability to command the loyalty of a veritable army of disaffected men who had been trained to the business of close-quarter fighting in the *Sturmabteilungen* of 1917 and 1918: the *Stosstruppen* (stormtroopers). Always as ready with boots and fists as with bombs, bayonets and machine-guns, these men transferred their fighting skills from the trenches of France to the towns and cities of their own land, forming themselves into armed bands known as *Freikorps* to combat the rising strain of Communism which threatened to seize power in the economic ruins of Germany immediately after the war. Subsequently, Hitler supplied them with rhetoric to justify their innate hatred of a faction which they believed had sabotaged their country's chances of military victory, and then all he had to do was point them in the right direction, and they carried out his bidding mindlessly, just as good soldiers should.

And they *were* good soldiers, in the main, and in the strictly military sense there was no doubt about that; during the war which engulfed Europe in 1939 they and their sons proved themselves to be among the best the world had ever seen: tough, resourceful, cruel fighters, generally well led, at both the strategic and the tactical level, and taking all before them. That they were not ultimately victorious was thanks largely to the fact that supplies (of both men and *matériel*) ran out before they had been able to secure their pivotal objectives, and then the inevitable followed.

There was another side to certain members (and indeed, to certain whole units) of the German armed forces, too, of course – an horrific side which displayed an atrocious disregard for even the most basic rights of others which amounted in the end to nothing short of genocide. In a sense, *this* was a form of combat, too – and it was certainly carried on at very close quarters – so it, too, falls within our ambit; we shall touch on aspects of it later, together with the brave attempts of partisan fighters to do what they could, often with minimal resources – which inevitably guaranteed that the struggle would be carried on at very close quarters indeed – to prevent the Nazi machine from swallowing their cultures whole.

The bellicose character of the German soldiers of the 1939–45 war was created very directly by the spirit of the men of 1918; men trained to assault enemy positions and take them by virtue of their ability to fight well – overwhelmingly well – at close quarters.

Italy – just as 'new' a nation as Germany; both were only unified in the 1860s – had never really found an external identity, despite more-or-less successful attempts to wrest segments of the Ottoman Empire away from Turkey before the First World War. It joined in that conflict on the Allied side, in an attempt to push back the borders of the Austro-Hungarian Empire, but succumbed to similar pressure applied less skilfully by an obscure ex-socialist journalist, Benito Mussolini, whose Blackshirts also introduced strong-arm tactics to domestic politics. The Italian *Fasci del Combattimento* never quite had the ruthless savoir-faire of their colleagues-to-be in the *Nationalsozialistische Deutsche Arbeiter Partei*'s armed wing, the SA (*Sturmabteilung* – a direct homage, this, to the stormtroopers of the war just gone, who made up most of its membership) and the later SS (*Schutz Staffel*), but that didn't stop them murdering, maiming and beating with a will.

Throughout the decade after the peace of 1918, Germany's internal politics were marked by armed struggle in the form of streetfighting and assassination. From the first, the *Freikorps* showed their willingness to deal harshly with dissenters – in the first years they killed thousands of non-sympathizers: 1,200 in pitched battles in Berlin alone in March 1919, and several hundreds in Munich in May; more in putting down a French-financed, Polish-led rebellion in Silesia the following year, and perhaps as many in the Ruhr, though that resulted in a 100,000-strong workers' rising which saw the armed bands temporarily cleared out of that region. With the rise of the Nazi Party, its armed factions proved willing successors to the *Freikorps*, for Hitler fully understood the power of terror and just how to go about instilling it in the people: by an overwhelming barrage of kicks and punches backed up with clubs and guns. The methods and manners of close-quarter battle had come to twentieth-century politics. Eventually, the *Schutz Staffel* (literally, 'protection squad'; Hermann Goering is said to have named it, after the defensive patrols the German Air Force had maintained above its airfields during the First World War) expanded into a second German army, in parallel with the *Wehrmacht* and often over-shadowing it, but always separate, and separately accountable. At its strongest, the *Waffen-SS* amounted to some 38 divisions, each nominally of 19,000 men, many of them volunteers from outside Germany.

The *blitzkrieg* strategy, and the earliest application of the tactics of what we now call the air–land battle, seem, on the surface at least, to be entirely at odds with the concept of close-quarter fighting; after all, as we have noted, the weapons which were central to that sort of combat had been developed precisely to enlarge the battlefield, depersonalize killing and take the war to the enemy before he was prepared to engage his own forces. But the two seemingly very different methods of fighting actually share an essence of ruthlessness all-important if one is to win. Hitler and his generals took that element of ruthlessness, that thrusting aggression which had carried the Nazi Party to power on the streets and in the beer halls, and converted it into a strategy for winning whole countries by means of a series of crippling blows to the head and body. First Poland succumbed, then Norway and Denmark, the Low Countries and France, then much of eastern and all of southern Europe (save for the neutrals in Iberia), as far as the Black Sea and almost to the shores of the Caspian, while simultaneously the *Wehrmacht* took most of North Africa and started a vast pincer movement which would, had it gone unchecked, have nipped off the entire Mediterranean basin. These were the sudden, all-out attack tactics of the hardened hand-to-hand fighter taken to an entirely new dimension and applied on a continental scale.

The stormtrooper tactics applied so successfully in March 1918, by von Hutier in the Somme sector (though usually attributed to him by name, they were actually developed at the OHL), were modified during the 1930s, both to take account of the newly developed ability of the *Luftwaffe* to supply accurate close air support and to incorporate one feature the Germany Army had previously disdained: armour. Up until the end of the Great War, the Germans constructed only a score of tanks of their own, the box-like 33-tonne A7V; they also made use of captured British tanks, but never had enough vehicles even to begin to need to formulate a strategy for using them; instead they threw them into the battle in an ad hoc fashion. There was only one tank-versus-tank battle during the Great War, near Villers-Brétonneux on the Somme front, on 24 April 1918, when fifteen German tanks confronted British Mark IVs and badly damaged two 'females' (machine-gun-armed tanks, for use against infantry) before being driven off by a 'male' Mark IV (a tank armed with a modified six-pounder field gun, firing solid shot).

The whole situation changed in the 1930s, with the introduction of the Panzer I in 1933. Though they had not produced armoured vehicles in great numbers themselves during the First World War, the German Army had not ignored the ability of the tank to plough through infantry-proof defences as if they were not there, and the new assault tactics incorporated the armoured spearhead concept for the first time, using tanks to suppress strong points and allowing the infantry stormtroops to force an infiltration. Simultaneously, the *Luftwaffe* was used to extend the reach of conventional artillery, employing purpose-built ground attack aircraft (predominantly the infamous Junkers Ju 87 'Stuka' – from *Sturzkampfflugzeug* – dive bomber) to bombard enemy installations just ahead of the attacking wave.

German forces even had the opportunity for a dress rehearsal for the Second World War, thoughtfully provided in July 1936 by the rightist faction in Spain. Determined to overthrow the elected Communist government, the rebels, led by Francisco Franco, Emilio Mola and José Sanjurjo, had the troops – in Morocco – but lacked the means to deliver them to where they were needed. The logistical gap was filled by the newly emerged *Luftwaffe*, whose Junkers Ju 52 tactical transport aircraft delivered 15,000 men of Franco's Army of Africa to Sevilla (while the Italian Regie Aeronautica covered the transport of a further large contingent by sea) and plunged the country into civil war. The first city seized by the rightists was Badajoz, no stranger to that particular act, though this time there was not the rape, looting and carnage which had characterized its seizure by Wellington's troops, a century and a quarter earlier.

The Spanish Civil War was to become notable for the foreigners who fought on both sides, some despatched by their governments (Germans and Italians, who fought with Franco, and Russians, who fought with the Republicans) and some volunteers, most – but by no means all – of whom sided with the leftists. At the beginning, however, it was a purely local affair between the rightist rebels on one hand and indigenous anarchists and Communists on the other. The rebels had arms and to some degree discipline – though as we shall see, it was sorely tried; while the workers had raw courage.

Republican Spain was a *cause célèbre* in the late 1930s, with supporters of the anti-fascist cause flocking to Spain to join the ranks of the Republic's International Brigade in their thousands. The Brigade started off with only French, Polish and German battalions, but grew to include contingents from most European countries and as far away as North America and Australia.

As with all Republican forces, the Internationals were poorly armed and trained. Their weapons came from captured Nationalist arsenals, arms smuggling or the Soviet Union. Bolt-action rifles and belt-fed machine-guns were the main weapons used by the Republicans, but a few sub-machine-guns did make their way to the Internationals. Machine-guns included French Hotchkiss and Chatellerault M1924s, British Vickers .303in and Lewis guns, Soviet SG43s and Degtyaryevs. French Lebel Modèle 1886, German Gewehr 98s, British SMLEs and Soviet Mosin-Nagant M1944s were the main rifles used by the Internationals. German Bergmann MP28s, Czech ZK383s, Soviet PPD 34/38s and American Thompson sub-machine-guns were much in demand with the Internationals. Not surprisingly, the wide variety of weapons in use with the Internationals played havoc regarding ammunition resupply throughout the war.

What of tactics? Although a few of the Internationals were veterans of the First World War, most had no combat experience and had to learn their military skills from scratch. They were thrown into battle with almost no training, and those who survived considered themselves lucky. Spanish Civil War infantry tactics put an emphasis on close-quarter action because most combat took place in cities or on mountain ranges.

The machine-gun and hand grenade were the weapons of choice in urban warfare. Machine-guns were sited in upper-storey windows to shoot along streets to stop advances. A single, well-sighted machine-gun could hold up a whole Nationalist *banderia* for days. Ammunition resupply was the key to keeping machine-guns in action, and both sides regularly used civilians to bring up extra ammunition.

House-to-house fighting was a bloody business. For defence, city apartment blocks were turned into fortresses, with doors and windows barricaded, roofs reinforced and cellars turned into bomb shelters. When assault parties tried to gain entry, grenades would be dropped on them from upper-storey windows or thrown into rooms just occupied by the enemy. If the enemy gained a foothold in a house or a neighbourhood, then it was vital to launch an immediate counter-attack to stop them consolidating their hold. In many street battles, whole districts of cities changed hands frequently as both sides battled for supremacy.

In mountain warfare, the defenders again had the advantage because of the machine-gun. Frontal assaults on trenches were as futile as in the First World War. Only artillery and tanks could break strong defensive positions. The fighting was merciless, particularly in the cities; those centres of industry and trade, such as Barcelona:

When, on that Sunday morning [in late September 1936], the Barcelona rebels marched out of their barracks against a sleeping population, they were met by hordes of workers, the mob, half dressed, unarmed or badly armed, called out of their houses by some strange insurrectionist telepathy. The military opened fire with rifles, machine-guns and artillery, and that mob charged them with knives, sticks, stones, little automatic pistols and here and there a rifle. Or charged them with nothing at all but fists and teeth and wiped them out . . .

Make this drama for yourself. The Paseo de Colon is a long avenue of date palms, on one side of which are the dock warehouses and the railways serving them. On the other, shops, ship chandling and brokerage firms, public offices, the Post Office and the military headquarters. In front of the headquarters and firing down the long avenue is a garrison of about five hundred fully armed soldiers; about twelve machine-guns, three or four pieces of artillery of between three- and four-inch calibre. The charge begins well down the avenue, the workers are creeping along inside the walls that shut off the railway from the avenue, firing through the railings. A few run along the warehouse roofs, one falls shot dead on to a wagon of tallow barrels. Suddenly the workers burst through a drive-in and emerge among the palms. All the rifles, machine-guns *and artillery* open fire, at once. The crowd goes roaring up the avenue, hundreds falling, the shells tearing lanes through them, bullets slashing the trunks of palms, smashing windows and chipping stone as the machine-guns spray.

They reach the military lines and unarmed men leap on the gunners, wrestle with them, strangle them, drag them to the ground and stab them with knives. Men dive at the machine-guns like football players and upset them with their hands, kicking, cursing, and tearing with nails, hammering out the brains of soldiers with lumps of paving stone. The soldiers break, but before they can withdraw, groups of workers drag the artillery around and at ten yards' range a shell smashes the headquarters gate. Others, not knowing how to aim, send their shells flying high into the façade, smashing a pillar, blowing in a cornice. The mob pours into the barracks with captured arms, and one docker and his mate, carrying a machine-gun, stagger up the stairs, firing from the chest. In another wing armed soldiers are trapped in a room and four anarchists blaze at them from open doors; the two survivors race along the corridor and open fire with a machine-gun across the quadrangle.

Suddenly, silence. Within a minute the mob is rushing the abandoned artillery to the central square, the Plaza de Cataluna, where the military have fortified themselves in the Columbus Hotel; men and women, at last with us, are already streaming out of the side-streets; a youth is pressed against the belly of a nude statue, his forearm is on its breast, he opens fire with a tiny pistol; another is stretched beside the fountain basin with a cheap revolver. The moment the workers arrive they rush the hotel; again the same drama is played, scores fall; but now they have arms and bombs. The door collapses, and the bedrooms, saloons and corridors are choked with frenzied men, swinging at Fascists with chairs, bars of iron, knives and rifle butts. Fascists are flung bodily out of windows, or down the lift shafts, or driven into a lavatory; a bomb is pitched in after them.

. . . By two of the afternoon, with perhaps twelve hundred killed and wounded, the city is quiet, in the hands of the workers.[1]

The idealists who came flocking to the workers' cause soon discovered that war – and especially civil war – is a very dirty business, but the presence among them of some of the finest writers and thinkers of their generation assured, at least, that the war would go down in the annals of history in such style and detail as few ever did before, or have since; the incidents recorded on the spot and the records themselves despatched to Paris, London and New York, where they were blazoned across newspapers and magazines for all to see. The following incident, seen through the eyes of volunteer Jack Roberts, took place at Villanueva de la Cañada on 6 July 1937:

It was dusk of the first day, and our Battalion (of the XV International Brigade) had advanced to within about 3,000 metres of Villanueva de la Cañada. Some were lying down in the ditch beside the road (to Brunete); five of us were taking cover behind the dung-heap on the right. Suddenly, someone shouted: 'Don't fire; there are children coming from the village. We looked down the road and saw about twenty-five people, men, women and children. In front was a little girl of about ten years of age. Behind her came an elderly woman, a boy of fourteen or so, some old men. The remainder were young men. As they approached, they shouted 'Camaradas! Camaradas!'

Believing them to be refugees, we answered them and called them forward. Some of us were now standing, some walking to meet and welcome them.

Pat Murphy, of Cardiff, was the nearest to them. He approached them, telling them to lay down their arms if they had any. For answer, a revolver blazed (though not at Murphy; he was wounded – in the groin – but later, by a grenade). Then the Fascists, who had been driving this group of old men, women and children as cover, started throwing hand grenades in our midst. For a few minutes, pandemonium reigned.

It was hard to distinguish friend from foe. The Fascists had difficulty also. I saw a figure bend down, jerk up a wounded man, saw the flash of a revolver, as the Fascist despatched one of our men. 'Commandante!' yelled someone, and Fred Copeman called out in answer. A Fascist lobbed a grenade towards him. It missed Fred, but killed Tommy Gibbons, Battalion Secretary. We had fallen for an old trick.

The crash of grenades, the barking of guns, and the shrieks of the women and children are still in my ears. But, in ten minutes it was all over; the last of the Fascists lay dead. Then, forward we charged and stormed the village. From the other end, the Dimitrovs [the Russian volunteer Brigade] came in. We lighted the streets with the red light of bursting grenades as we drove the Fascists before us to the centre. The Dimitrovs and ourselves were within an ace of charging each other [but] luckily we were shouting anti-Fascist slogans, and recognized one another in time.

The Spanish Carabineros, great fighters, were also in action, helping us. We mopped up, street by street, men fell. We succoured the wounded, no matter whether they were ours or the enemy. Bill Meredith, brave young Company Commander who had been a hero at Jarama and throughout the day, bent down to attend to the cry for help. For answer he got a bullet through the heart from the wounded Fascist.

Later, while going round to identify those of our comrades who had fallen, we could not help thinking of the brutality of Fascism when we found the bodies of the little girl, the elderly woman and two elderly men.[2]

Some volunteers discovered, too, that there is more to war than simply fighting, and that even humanitarian acts sometimes require great courage. Joe M. Gordon, one of many Americans to volunteer, fought with 1 Company, Lincoln Brigade. His company commander, John Scott, was shot in the head by a rebel sniper; the incident took place at night, during the final days of February 1937:

... Scott was lying flat on his stomach with his right arm under him, his head twisted sideways. Bill Henry was pushing dirt in front of Scott's head to give him some protection. At every move he made he drew fire ... When I asked him how far the Fascists' lines were, he told me sixty metres ...

I moved over to Scott ... With his left hand he took one of mine. I could feel his strength slowly ebbing away. I told him I would go back and bring aid. He squeezed my hand hard and said, 'Don't do it. It's a waste of time.' 'What the hell do you mean, it's a waste of time? You're a human being, ain't you?' With all the suffering he was going through, a smile came over his face. I hated to break hands with Scott. It seemed as though I was giving him strength through my hand.[3]

Gordon reached the Aid Station without mishap, and eventually was able to obtain a stretcher and three men to help him, and set off back toward the fighting.

After what seemed hours, we finally came to Scott. We grabbed him none too gently – we couldn't help it – and put him on the stretcher. He was groaning slightly; he couldn't groan any harder if he wanted to, he was so weak ...

Now the question of how to get back to the Aid Station – if we were to crawl along in the dirt and mud, or go by the road. We decided to go by way of the road even though it was more dangerous. Four men, including Burns and Shappiro, grabbed the [stretcher] handles, lying flat on their backs, counting three, then up and backwards, then digging your feet in the dirt push your way back to position [*sic*]. Poor Scott, what a target! It was a good thing he didn't know what was going on. After what seemed ages, 150 yards all told, we finally reached close to the road. I immediately hopped off the embankment, grabbed the two handles of the stretcher and gave a hard pull, just as the Fascists opened up terrific fire right on us. Everyone was wounded except myself . . . What a hell of a situation! You go after one wounded comrade, and now look at the mess![4]

Gordon set off for the Aid Station again . . .

I started crawling on the side of the road. About three minutes later a terrific barrage opened up. Not moving, lying flat on my face, I was hoping the fire would subside a little, so as I could move on, but it seemed to get heavier . . . I decided to push on, knowing that if I stood [*sic*] in the same spot, sooner or later I'd get it. Pushing myself with my feet and using my hands, not daring to raise my body, I moved forward slowly. I got a cramp in my left leg, also started to vomit. Resting a few minutes then continuing onward, I finally came in sight of the dead comrade. Crawling up to him, I fixed his body so as to give me as much protection as possible. Soaked with his blood which continued running, I don't know how long I lay. I'm sure he saved my life. I was almost afraid to breathe lest I sniff in a bullet.[5]

Gordon reached the Aid Station, and reported the situation to his Battalion Secretary, Cooperman. After an abortive attempt to set off up the road with an ambulance, the two, with volunteers named Greenleaf and Tanz (and one other) together with a first aid man named Toplianos, took another stretcher and set out again, passing two of the original rescue team who had made their own way back despite leg wounds:

We started up the road, almost to the half-way mark when we bunk [*sic*] into Paul Burns and another comrade, Gomez, carrying in Scott. Telling the other comrades to wait for us, Toplianos and I carried Scott into the Aid Station. He was still alive.

Toplianos and I went up the road again and caught up with the others. Comrade Tanz was not around. We got off the road and started crawling, but not too far from the road because we knew if there was anybody left he would still be on the road. The Fascists opened up bursts of fire. Whenever they opened up we would be still and then continue. After a hard burst of fire at us had ended, I pushed the comrade next to me and said: 'Let's go!' He didn't move. I looked at him and found that comrade Ralph Greenleaf had been shot right through the helmet. He died instantly, without a sign, a pool of blood forming quickly.

We continued pushing up. Hearing groans, we stopped, called softly: 'Where are you?' No answer, except the monotonous groans. We could tell he was nearby. Toplianos finally spied him. He and I climbed down the bank. When we got there, he was lying in the middle of the road, groaning very loudly. He [it was Shappiro] was in terrible pain from an explosive bullet wound in the ankle.

He was very, very heavy. We got him to the embankment, where he was grabbed by two of the comrades above, and all together we lifted him off the road, onto the dirt, and got him on the stretcher.

Again we started back in the same manner we had taken with Scott. After each yard we had to rest, he was very heavy, dead weight. His ankle or foot kept on turning around, and he kept groaning terribly, drawing fire. After what seemed a lifetime, we finally got him back into the Aid Station. What a night! Killing is a pleasure compared to the saving of life.[6]

The Spanish rightists were well known, even before the war started, for their straightforward attitude towards Communists: they were to be shown no mercy, and if possible, simply shot out of hand. One incident involved a controversial Englishman named Richard Meinertzhagen, who played a very shady role indeed throughout much of his long life. (He always claimed to have masterminded the mission which spirited the young Romanov princess, Anastasia, away from Ekaterinburg in 1918, before the rest of the Russian royal family was murdered, though there must be grave doubts as to whether the incident ever took place. That entire story actually makes better fiction than fact, as evinced by author Brian Innes's treatment of it in *The Red Baron Lives*.) Among Meinertzhagen's other activities was the part he played in trying to suppress the Soviet Union's influence in Spain before the civil war:

I have had to come to Spain [he wrote in his diary for 1930; like much of Meinertzhagen's life story, his diaries do not always bear the closest scrutiny, there being substantial allegations that he 'fudged' his entries later, as new facts came to light, as well as doubts as to his overall veracity] to try to terminate an intolerable situation and once and for all put a stop to the activities of Russian revolutionary activities [*sic*] in western Europe and incidentally to my own personal security, a matter which I seldom speak about and have so far never mentioned in my diary. It is a secret well kept and know [*sic*] to a very few, that my main work during the past seven years has been the hunting down Bolshevic agents in western Europe and keeping in touch with their activities. It has been exciting enough, profoundly interesting and not without a spice of danger . . .[7]

Meinertzhagen's biographer, Mark Cocker, his tongue fairly firmly in his cheek, describes how a gang of Soviet agents had been dislodged from their Amsterdam base and moved to Spain, where they took up residence in a large, isolated house some kilometres from Ronda, a small mountain town in Andalusia. Meinertzhagen, being 'more conversant with the organization than probably any other man in Europe' (at least, by his own account), was asked by the Spanish authorities to co-ordinate and effect their arrest. With a team of Spanish policemen, he gained entry into the house on the night of 3 March (his 52nd birthday), managing to handcuff and gag five of the seventeen Russians before one of them sounded the alarm. What followed was straight out of the sort of adventure fiction which was so popular at the time but was a factual forerunner of much of the urban combat during the Second World War, when towns and cities were cleared of enemy troops house by house and room by room:

Would they fight? If so it meant sudden death for many or all. Stegmann came out first, pistol in hand, a black-bearded giant. I shouted 'hands up' but it was too late. He fired at me. I recollect the feeling of satisfaction like a flash, that he had fired first, giving us a free hand and putting himself and his whole gang in the wrong [this, of course, is one of the essential differences between police work and military action]. I fired at once bringing him down with an angry snarl and not before he got his pistol off again. He

67

was like a wild animal on the ground with ears back, teeth gleaming and eyes ablaze, though knowing death was fast overcoming him. It was horrible to see.

Two of my men were down and apparently dead. From that moment there was a scrimmage. Doors seemed to open in all directions, pistol shots rang out from every corner and angry shouts drowned any orders I was able to give. I yelled to my men to keep together, but they would not. The world seemed to revert to the jungle and jungle law. Social man became a savage beast. All our worst instincts surged up and I saw red, realizing that it now meant a fight to the finish, so I let fly in every direction, placing bullet after bullet into those wretched Russians. All the vaneer [sic] of civilization went in a flash, to be replaced by stealth, crouch, cat quickness, lightning spring and the steel claw which rips the jugular vein. The landing was soon a shambles and for a moment I thought we had got the worst of it, but quite suddenly everything stopped except for defiant curses from one room. Our blood was up and we at once took up the challenge and rushed in to despatch what remained of these men. Another man [went] down before we broke down their resistance and then someone threw two small bombs which exploded with apparently harmless results but in a moment their meaning was realized. I caught my breath as I smelt a burning gas. One gulp I swallowed before I realized the poison. I yelled to my men to get out and we hurried away dragging six dead and wounded with us.[8]

The reader is left to form his own opinion of the credibility of this account; this author, frankly, tends to disbelieve it. Yet this *was* the same Richard Meinertzhagen of whom Elspeth Huxley, in a preface to a published version of a portion of his diaries, said: 'Richard Meinertzhagen was a killer. He killed abundantly and he killed for pleasure'[6] and it *was* the same Richard Meinertzhagen who, as a young colonial officer, ordered a machine-gun turned on a fifty-strong native band at Kaidparak Hill in Kenya in 1902, killing twenty-three outright (though one must admit, behaviour such as that was by no means unknown in Africa at that time).

But there were worse than Richard Meinertzhagen abroad in Spain in the 1930s. On 26 April, of the second year of the war proper, 1937, having already participated in combat on a small scale, the *Luftwaffe*'s Kondor Legion, under General Hugo Sperle, was directed to test out its capacity for large-scale destruction in a co-ordinated attack on the Basque town of Guernica. After a three-hour bombardment (some aircraft – the original Ju 52s, augmented by newer Heinkel He 111s and Dornier Do 17s – flew more than one sortie), 1,654 people lay dead and a further 889 were wounded. German Air Force chief Hermann Goering, himself an 'ace' from the First World War, later (in 1946, at the time of his trial for a multitude of war crimes) admitted that the raid was mounted simply to test the new German air warfare strategy.

Thus Germany tested two quite separate untried elements essential to the concept of the air–land battle under more or less realistic conditions: they transported large bodies of infantry by air, and began concentrated aerial bombardment on a large scale. The latter, one should perhaps point out, was untried in Europe, for one can virtually exclude forays made in the field by both sides during the Great War, even though the best part of 2,000 Londoners did die in air raids; it was being practised at that same time by the Italians in Ethiopia and more widely by the Japanese in their partially successful attempt to overrun China, particularly at Nanking. The theorists – chiefly Douhet in Italy, Mitchell in the USA and Trenchard in Britain – maintained that wars could actually be won by these means alone, but have never been proved correct, even when a later technology had taken most of the guesswork out of bomb-aiming.

There remained a further refinement of the first element to be practised: airborne assault, either by parachute or by means of disposable aircraft, usually unpowered gliders. This approach was the brainchild of General Billy Mitchell, too, though probably not uniquely; as early as 1918, Mitchell had advocated to the American commander in Europe, 'Black Jack' Pershing, that part of the US 1st Division be transported by air to stage a parachute assault on Metz, behind the German lines. Pershing vetoed the idea as impractical, which it probably was, given the state of the technology at the time. The first practical military jump was made in Italy in 1927, and by the late 1930s France, Germany, Italy, the USA and the USSR had all experimented with airborne operations. It was the Red Army under Tukhachevsky (one of the first military officers to be 'purged' by Stalin; he was executed in June 1937) which created the first airborne brigade, in 1932, and the following year a combined force of 7,000 men, parachutists and air-mobile, were deployed during an exercise near Minsk. By 1938, the Red Army had four airborne brigades, and the *Luftwaffe*, having followed developments closely, had formed an airborne division consisting of nine battalions, seven of them air-mobile and two parachutist. From then on, airborne assault became a regular if not constant feature of many tactical deployment plans, while the paratroopers themselves rapidly assumed elite status. Due to the way they were deployed, airborne troops frequently went into action with little or nothing in the way of heavy weaponry and, not unnaturally, much of their training was concentrated on close-quarter combat; hence they will occupy us considerably from here on.

From the outset, the bodies of men specifically and particularly charged with close-quarter battle, starting with the original Sturm-abteilung Kaslow and Rohr, raised from the German Army's 18th Pioneer Battalion in March 1915, through the *Stosstruppen* of 1917 and the air-mobile and amphibious 'commandos' of the Second World War and later, to the so-called 'Special Forces' of the postwar period, have earned their elite status by virtue not so much of the evidently dangerous nature of their role – all soldiering is dangerous, and as we have previously noted, a bullet from a thousand metres (or an artillery shell from thirty thousand) kills just as effectively as a knife or a garotte – as by their willingness to dare. Not for nothing do the men of Britain's Special Air Service Regiment have 'Who Dares Wins' as their motto; and the competition in question is not just the fight of the moment, but for peer group recognition as the best in their chosen field.

Chapter Three

Blitzkrieg!

The main difference in character between the Second World War and the war which had almost devastated Europe two decades earlier was the degree of mechanization employed, with aircraft and armoured vehicles finally fulfilling the promise they had shown and ensuring that there would be no return to permanent static combat (though there were to be exceptions, particularly in areas where armour couldn't operate). Just as the repeating rifle and machine-gun had lengthened the battlefields of the First World War, so the new weapons carriers seemed set to expand them still further, with new and more powerful (and, crucially, more accurate) artillery also contributing to what was to be the first general 'over the horizon' war. Yet, just as close-quarter fighting had continued to play an important part in deciding the First World War, notwithstanding the fact that by far and away the greatest number of its casualties were caused by artillery and machine-gun fire, so it was also to prove to be a vitally important component of the Second, as it became clear to the generals that while the new weapons could certainly assist (at least) in the taking of ground, only infantry could hold it, if necessary by means of hand-to-hand fighting.

The *Waffen-SS* and *Wehrmacht*, ably assisted by the Luftwaffe, began putting the invaluable lessons they had learned so cheaply in Spain into practice on Germany's own account in September 1939, ripping through southern and western Poland with a speed and ferocity unmatched in modern war, while Stalin's Red Army stole the other half of the country under the guise of protecting its own borders. Britain and France mobilized, but stood impotently by. Then the war ground to a temporary halt, but only over the winter. In April 1940 Hitler invaded Denmark and mineral-rich Norway, and the next month it was the turn of Belgium, Luxembourg and The Netherlands, all of them neutral but providing an easy way for the *Wehrmacht* to circumvent the defensive Maginot Line and advance on the bitterest enemy of all, France. It was 1870 and 1914 all over again, but this time with infinitely more horrific results.

From the very beginning, the *Waffen-SS* divisions' training and aggressive spirit, in particular, proved quite as solid as its commanders had believed they would, and those two factors had a telling effect in battle. For young soldiers like Sepp Lainer, an Austrian who had joined the elite SS Der Führer regiment in the spring of 1938, the experience was both exhilarating and terrifying. Having missed the Polish campaign, Der Führer – now part of the SS Panzer (i.e., armoured) Division Das Reich – spearheaded the drive into Holland, attacking the fixed defensive strong points the diminutive Dutch Army had built all over the country; the offensive went absolutely according to plan, both overall and on a local level:

They reached the field of fire of the first bunker, which was flanked by two machine-guns. 'Two volunteers to take out those two machine-guns,' called Hauptsturmführer Harmel. To his own amazement, Sepp Lainer stepped forward. His friend Weickl did the same.

'All right, Lainer, go round to the left. When you are within certain hand grenade range, take out the machine-gun nest with three grenades. Weickl, you do the same on the bunker's right flank. Take the two communication trenches, left and right. Be careful, there might be a few Dutchmen hiding there.' Harmel looked at his watch. 'Throw your grenades in five minutes. Afterwards we will attack and open fire, so keep your heads down.'

Sepp Lainer slipped through the chest-high trench until it turned and led diagonally towards the machine-gun. There he would have to leave the trench, because there would be nothing between him and the enemy troops in the machine-gun nest. If he made the slightest noise they would spot him and he wouldn't stand a chance. Lainer left the trench and moved forward through some low bushes. He only half-listened to the firing of his comrades and the enemy as he concentrated on the machine gun nest. He took cover briefly when a burst of machine-gun fire hissed close by, but it was obviously meant for his comrades further back in the trench. He crawled another ten metres and saw one of the

Dutch soldiers. The man was looking over the parapet, but not in his direction.

The shallow pit in which he was lying allowed Lainer to throw his grenades from a kneeling position. In this way it was possible for him to toss the grenades a good thirty metres. He looked at his watch and saw that he still had thirty seconds to go. He laid two of the grenades in front of him on the edge of the pit, ready to throw, screwed off the safety cap of the third, gripped the fuse's porcelain knob between his index and middle fingers and waited. When the minute was up he pulled the cord, waited two seconds and threw the first grenade into the enemy machine-gun nest. Even before it detonated he had thrown the second. As he was throwing the third grenade, the first exploded. Through the noise of battle he heard more grenades exploding to the right; that had to be Weickl. Behind him he heard his comrades open fire and through the noise of battle he could hear the tramp of their running boots. A man landed beside him in the trench; it was the company commander. 'Well done, Lainer,' he called. 'Now let's get that bunker . . .!'

When it was finally all over, Sepp Lainer found that his hands were shaking uncontrollably. He hid them behind his back, but a glance at the others was enough for him to see that they were experiencing the same thing. The battalion commander, who had been with the 3rd Company, came over to them. 'Well done, men! That's the kind of assault I like!' he cried, and handed out several packets of cigarettes.[1]

The keynote of the German offensive as it moved on into France was its speed, and for the individual infantrymen, that meant that small pockets of stubborn defenders, particularly those guarding key strategic objectives such as bridges, had to be assaulted immediately with whatever resources were available locally; there could be no waiting for heavy weapons support:

Rudi Brasche felt his hands, which were gripping the handles of the ammunition boxes, becoming wet with sweat. He was afraid, but he knew that he would have to overcome his fear to be effective. The pace quickened. Brasche had to keep pace in case Nehring should need ammunition. Richard Gambietz paused and turned toward Brasche. His young friend tried to smile, but only managed a grimace. Brasche caught up with him and moments later they joined Nehring. Just then they heard the roar of an engine, and the pointed nose of a French armoured car pushed its way through the thicket. They all saw the machine-gun and the 20mm cannon.

Feldwebel Wegener signalled silently as Brasche looked his way. He raised one of the four hand grenades he had stuck in his belt. Gambietz nudged Brasche, and pulled out a grenade. Brasche put down the two ammunition boxes. He screwed off the metal caps of two grenades and crouched lower in the shadow of the bushes as the armoured car rolled directly towards him.

The vehicle came nearer and nearer. Then Brasche spotted movement behind it: enemy infantry, trying to approach the bridge from the flank under the protection of the armoured car. The French vehicle was now thirty metres away. Glancing to the side, Brasche saw that Nehring had aimed his machine-gun at the enemy troops.

'Go!' shouted the Feldwebel. Rudi Brasche jumped up and threw one of the grenades. Gambietz and the Feldwebel did likewise. The enemy were already firing back when the grenades fell on the armoured car. Nehring opened fire with his machine-gun. The three hand grenades exploded. Brasche jumped to his feet again. A bullet whizzed past his head. Behind him, Nehring's machine-gun roared again. Three figures appeared in front of him. He ripped the porcelain knob from the grenade and tossed it towards the group of three. At the same time he threw himself down head first to the ground. The burst of fire meant for him passed overhead. The blast from the hand grenade silenced the three enemy soldiers.

On the far side Feldwebel Wegener reached the second armoured car and threw two grenades inside. Gambietz overcame a second group of enemy troops. Then it was quiet except for the sound of breaking branches as the enemy fled through the woods. There was a fresh outbreak of firing to the right and then from the far side of the bridge, where a squad had been left behind.

Three armoured cars made another attempt to reach the bridge, but were forced to turn away. Then Oberstleutnant Oskar Radwan, commander of II Battalion, arrived, at the head of the relief forces. He was the first to roll across the bridge and into the enemy. As had often been the case before, this bold move succeeded and the area in front of the bridge was swept clear.

They had captured the Seine river crossing undamaged. The advance road lay open before them.[2]

In no time at all, it seemed, France was driven to her knees, and the British Expeditionary Force which had been sent there to stiffen the defences in the north (shades of 1914, but this time without superior infantry tactics and individual firepower) was driven into the sea off Dunkirk and became the subject of a desperate, all-out attempt to save at least its men to fight another day, even if almost all their equipment had to be left behind. By the time the war was eighteen months old, Hitler was master of all the land from the Black Sea to the Baltic, controlled much of North Africa and was poised to strike at Russia. Britain (and once again, and henceforth through the period of the Second World War, we include the hundreds of thousands of Empire troops fighting alongside the British here) opposed him virtually alone, aided by a sprinkling of Belgians, Czechs, Danes, Dutchmen, Frenchmen, Norwegians and Poles who had managed to escape when their own countries were overrun.

There was no battlefield closer than North Africa where the land war could be taken to the enemy, and there the campaign was not going well. Indeed, it looked, in early 1941, as if the fortress of Tobruk would become the site of a desperate Dunkirk-like evacuation, but instead its garrison, largely composed, initially, of Australians and Indians, fought on, with the strength and courage of desperate men, and held out for nine months in a static fight reminiscent of those of a quarter of a century earlier:

In Post 33, just west of the El Adem road, Lieutenant Mackell, a platoon commander, looked at his watch. It was eleven o'clock. Beside him, Corporal Jack Edmondson, a sheep farmer's son from New South Wales, peered into the darkness and the thinning dust cloud. He gave a low whistle.

'Jerries! A whole heap of the bastards!'

Mackell's eyes followed the corporal's pointing finger. Vaguely he could see the Germans moving through a gap in the wire to the east of his post; twenty or thirty of them.

'Let 'em have it!'

As the machine-guns opened up, to Mackell's astonishment the Germans replied with mortars and light field guns, plastering the post with shells. This was no patrol . . .!

75

By midnight the situation was fraught with danger, and it began to look as if the whole line had been breached and that soon the enemy armour would come rolling through the gap. Dug in about a hundred yards away, it was obvious that the Germans were determined to create a bridgehead through which their tanks could pass. It was equally obvious that they could not be driven out with small-arms fire. Mackell decided to do the job with bayonets and grenades.

With Edmondson and five others, covered by fire from the post, he sprinted into the darkness, running in a wide semicircle so as to approach the Germans from their flank. Almost immediately they were spotted, and all the enemy guns were turned on them, forcing them to go to ground.

A second dash brought the party to within fifty yards of their objective. Once again they threw themselves down, and as they lay there panting, pulled pins from their grenades. Then, with a wild yell, they charged.

For a few terrible moments the seven men came under a deluge of fire. Edmondson took a full blast of machine-gun fire in his belly just as he hurled his grenade into the midst of its crew. Another bullet hit him in the throat, but he kept on running and, having thrown a second grenade, went in among the Germans with his bayonet. In a panic of terror they bolted out of their trench, running blindly into the wire, to hang screaming there as their attackers bayoneted them.

Soon Mackell was in difficulties, for as he fought with one German on the ground, another came for him with a pistol. He yelled to Edmondson for help. With blood pouring from his throat, the corporal dashed to his rescue, killing both Germans with his bayonet. In the hand-to-hand fighting which followed, Mackell, having broken his bayonet in a German's chest, clubbed another to death with his rifle butt, while Edmondson and the others killed a dozen more. The corporal went on fighting until the last of the enemy had fled, abandoning their weapons, and then collapsed. The others carried him back to Post 33, where he died, soon after dawn. Corporal Jack Edmondson was awarded the Victoria Cross.[3]

Patrolling was an essential activity at Tobruk, too, in order to maintain an up-to-date intelligence assessment, as well as to keep the enemy at full stretch, and the Indian Rajputs serving with the

British Army proved to be supremely good at it, never returning without claiming a high score of Germans and Italians killed. Anthony Heckstall-Smith (from whose *Tobruk* the account of Corporal Edmondson's suicidal bravery is also taken) recounts that a Sikh regiment's commanding officer became suspicious that perhaps his men were overestimating the number of enemy dead, and lectured them sternly. A few nights later, he says, when the patrol returned through the wire just before dawn, its leader laid two small sacks before his CO, before saluting smartly and turning away. They contained irrefutable, if macabre, evidence of the success of the night's operation, for in one were sixteen right ears, and in the other, sixteen left ears, all of them still oozing the blood of their former owners.

One of the Sikhs' strongest qualities was their ability to move silently in the darkness, and thus to be able to take isolated enemy soldiers completely by surprise. And it wasn't just enemy soldiers they surprised, either: 'It was one of those nights,' a South African sapper major, clearing mines close to Sollum, told Heckstall-Smith,

> when the desert was as quiet as the grave. I was on my knees dealing with a detonator when suddenly a hand was clapped over my mouth and a thin arm wrapped itself around my neck. I was so numb from fear that I could not move ... I felt someone running their fingers over me, feeling the buttons on my tunic, my revolver, my pockets. When they reached the crowns on my shoulder, they seemed to hesitate. Then, just when I'd made up my mind that I was as good as dead, I was released. I heard a voice whispering, but I couldn't hear what was said because the blood singing in my ears deafened me. But I did catch the single word 'Sahib'. I don't know, but I may even have passed out in sheer terror. But when I pulled myself together, I was still kneeling on the ground, staring up at two bearded Sikhs. They were out on patrol, and since I was in the Jerry minefield, I'd had a pretty close shave. I still get the trembles every time I think of it![4]

While the British were fighting to maintain a tenuous toe-hold in the sands of the northern Sahara, their new-found allies in the Soviet Union were being driven steadily back through European Russia by the greatest blitzkrieg of them all, Operation Barbarossa, which burst on a largely unsuspecting Red Army in the opening hours of 22 June 1941. To begin with, the Soviet troops fell back quickly, and by the end of September, the German front extended from Leningrad to Odessa. On the Central and particularly on the Southern Fronts the Germans were to overrun a great deal of territory before being thrown back, but Leningrad itself never surrendered, and the fighting on the River Neva, to the south of the city, in the winter of 1941–42 was as fierce as any in the entire theatre, with both sides struggling for mastery of a line of opposing strong points:

'The Ivans are coming!' *Unteroffizier* Juschkat reached the entrance to the bunker he commanded, part of a strong point they called 'The Wasps' Nest'. Within seconds he had awakened his men. Moments later the heavy machine-gun [an MG34], which could sweep the forefield as far as the enemy-held hill, was manned. Pietsch and Gross, the two machine-gunners, peered through the embrasures, but there was nothing to be seen.

'Where is the light machine-gun?' Bichlap and Ochtrop, the two members of the LMG team, followed Juschkat outside. They went to the left end of the trench in front of the bunker and set up the gun where it could cover a blind spot.

Obergefreiter Paul held field-glasses pressed to his eyes and scanned the forefield, where scattered bushes and uprooted trees offered ideal cover. He spotted the first Russians when they were about 100 metres from the German trench and their white snow smocks emerged from the grey-white of the hollow. 'There they are, Franz!' he cried, and Juschkat suddenly saw several clouds of white mist: it was the exhalations of the Russian soldiers who had so skilfully approached the German positions. In the cold it gave away their location. Juschkat crept over to Bichlap.

'Target in sight!' he said, even before the *Unteroffizier* could speak. 'Open fire!' As the machine-gun began to rattle, Juschkat, too, opened fire with his rifle. He took aim at a Russian who had come into the hollow on skis. Exactly three seconds after the action began the heavy machine-gun opened fire; at the same time one of the Maxim guns the Russians had hauled into the depression [the Red Army still relied on the design the Imperial Russian forces had taken up around the turn of the century] also began to fire. Clusters of tracer from the Russian weapon hissed over the rim of the trench. The bullets struck the ice-covered bunker with a smacking sound. Finally, the two neighbouring strong points opened fire as well. The entire snow-covered plain in front of the Battalion came to life as the Russians jumped up and came running towards the German positions.

Three or four Maxims were firing now, and they were joined by other automatic weapons from the hill and two field positions on the left and right. The Soviets drew nearer and nearer. They were scarcely forty metres from the bunker, when the flanking machine-gun fell silent.

'Jammed!' screamed Bichlap, trying desperately to clear the stoppage. Juschkat dropped his rifle and pulled his 'Zero-eight' pistol from its holster. He released the safety with his thumb. With loud shouts of 'Urray!' the Russians charged the German lines on a broad front, the bursts of fire from the heavy machine-gun now passing harmlessly over their heads.

'Take the ones on the left!' roared Paul. The light crack of pistol shots was drowned out by the crash of exploding hand grenades, which Paul was lobbing towards the attackers. At the last minute the light machine-gun opened fire again and halted the onrushing Soviets. Franz Juschkat was amazed at how quickly the Russians could disappear. They were like ghosts. They concealed themselves behind snowdrifts and bushes and in depressions. They were still there, though, and the enemy soldiers fired on any German who allowed himself to be seen.

The men in 'The Wasps' Nest' tried to spy over the parapets. Paul crawled over to Juschkat, dragging a metal hand grenade box behind him. 'Here, Franz,' he said, and the two took out grenades, unscrewing the caps before placing them in a niche in the frozen trench wall, within easy reach. A dull, rumbling sound caused them to wince; they instinctively ducked down against the wall of the trench and there was a thunderous explosion as a satchel charge exploded a metre in front of the trench parapet. The shock wave forced the air from their lungs.

'Get up! Get up!' shouted Paul. Acting on instinct, Juschkat grabbed two hand grenades as he got to his feet. He saw the onrushing enemy troops about twenty metres in front, threw the two grenades and drew his pistol. The grenades tore gaps in the enemy ranks, but they were not enough to stop the assault. A dozen white-clad figures leapt into the trench. Juschkat saw a tall figure right in front of him. He fired as quickly as he could, and then a blow on the side of his helmet dropped him to his knees. In doing so, he avoided a burst of sub-machine-gun fire, which smacked into the trench wall.

Obergefreiter Paul grabbed up his rifle and swung it at the enemy soldiers. The men of the heavy machine-gun crew fired on the following Russians, further back in the forefield, and were soon joined by Bichlap on the flanking machine-gun. Juschkat inserted a fresh magazine into his pistol and eliminated the last of the intruders, while Paul went back to throwing hand grenades, and the attack bogged down. Not so to the left, where, around *Feldwebel* Korting's strong point could be heard the roaring of the Soviets. Above the noise of the enemy troops, shots rang out and hand grenades crashed.

'Mertens, Paul, Jagst, follow me!' They reached for their hand grenades and ran along the trench at a crouch. Three times they had to step over dead or wounded Soviet soldiers before they began to take fire. Juschkat saw the Soviets who had reached Korting's bunker. He threw two hand grenades in quick succession and then ran forward. His upper body was fully exposed to the Russian riflemen, but there was no firing from the forward Russian trenches, perhaps out of fear of hitting their own people.

The Russian infiltrators, caught in the act of slaughtering the machine-gun crew, were cut down before they could turn round, and in the lull which followed, Juschkat contacted battalion headquarters on the field telephone, alerting them to the situation and requesting support for the surviving members of the strong point's crew. Then he and his three-man relief force dressed the wounds of the injured men as best they could; the two men on the machine-gun were beyond help, their bodies completely riddled with bullets . . .[5]

That winter saw the entrance of the United States of America into the war, following Japan's surprise attack on Pearl Harbor and Hitler's even more surprising decision to declare war on the USA himself. With that, the conflict became truly global, and over the course of 1942 spread like wildfire from the Russian steppes to the coral atolls of the South Pacific. And throughout the year, and almost everywhere, it was the Axis forces of Germany and Japan which made the headway.

The garrison at Tobruk was relieved in November of 1941 during Operation Crusader, which saw British troops advance as far as the Gulf of Sirte. Success was short-lived, however, and, though short of supplies – much *matériel* having been diverted to the Eastern Front, where the war was now six months old – Rommel's

Afrika Korps (now renamed the Panzerarmee Afrika) struck back, and by the following May was once again in a position to dominate the coastal city. On 20 June 1942 Tobruk finally fell, this time after just two days of assault, and by the end of the month, Rommel had driven the British back almost within sight of Alexandria and the Suez Canal, arguably the most important strategic objective in the entire Mediterranean theatre of operations. The months which followed were the darkest of all for the Allies; worse, even, in global terms, than the summer of 1940, for by now Hitler's forces were almost at the Caspian Sea, with much of European Russia in their grasp, and had virtual control of the entire Mediterranean basin, either directly or through allies or more-or-less friendly neutrals. But the strain of fighting a war on such a vast front was beginning to tell . . .

The tide was stemmed and turned at El Alamein in the last week of October, in a battle which, just as General Montgomery had predicted, was 'a real rough house . . . a killing match', and one in which he refused to engage until his men were trained and hardened and equipment and supplies of *matériel* were built up to his satisfaction. First, there were gaps to be made through the mine-fields – perhaps the most bizarre close-quarter combat of all, against an enemy who was unseen but always at hand – to allow the passage of infantry and armour. Recalled Sapper John Jeffris:

Our three platoons were ordered to open three gaps in the enemy minefields so as when complete it would allow the armour through. My job was to walk into the minefield on foot laying white tape all the way, to mark the edge of the gap . . .

You never knew what type of mines were laid by the enemy. The German Teller mine was an anti-tank mine. It had a cast-metal case, and was shaped like a dinner plate. Their anti-personnel mines [S-Mines, which the Americans came to call 'Bouncing Betties'] were filled with quarter-inch ball bearings at the start of the war, and later on filled with any old scrap iron. If stepped on, they would jump three feet in the air, and then explode, spraying the whole area with shrapnel . . .

The detection party's job was to find the mines, either with detectors or by prodding with wire probes or bayonets [the procedure was to prod at an angle, and gently – hoping either to strike the non-reactive side surface of the mine, or to encounter the top surface, where the striker plate was positioned, with insufficient force to detonate the charge], and then mark them for the clearance parties. The clearance parties lifted the mines. You had to be careful, as the Teller mines often had booby traps fastened either to the side or to the bottom. So you had to feel all the way round to make sure.

What fun – I don't think.[6]

Then, to the accompaniment of the heaviest artillery barrage Africa has ever seen, Montgomery sent his Desert Rats into action. Lieutenant H.P. Samwell was a platoon commander with the Argyll and Sutherland Highlanders at El Alamein:

The tension [on 22 October] was almost unbearable and the day dragged terribly. I spent the time going over and over the plan of attack, memorizing the codes and studying the over-printed map which showed the enemy positions. Oddly enough, though keyed up, I did not feel any fear at this time, rather a feeling of being completely impersonal, as if I was waiting as a spectator for a great event in which I was not going to take any active part . . .

Oddly enough, I don't remember the actual start – one moment I was lying on my stomach on the open rocky desert, the next I was walking steadily as if out for an evening stroll . . . I suddenly discovered I was still carrying my ash stick. I had meant to leave it at the rear Company HQ with the CQMS and exchange it for a rifle. I smiled to myself to think that I was walking straight towards the enemy armed only with a .38 pistol and nine rounds of ammunition. Well, it was too late to do anything about it by that time, but I expected that someone would soon be hit and I could take his . . .

The line broke up into blobs of men all struggling together; my faithful batman was still trotting along beside me . . . I saw some men in a trench ahead of me. They were standing up with their hands above their heads screaming something that sounded like 'Mardray'. I remember thinking how dirty and ill-fitting their uniforms were. To my left and behind me some NCOs were rounding up prisoners and kicking them into some sort of formation. I waved my pistol at the men in front with their hands up to sign for them to join the others. In front of me a terrified Italian was running round with his hands above his head screaming at the top of his voice. The men I had signalled started to come out. Suddenly I heard a shout of 'Watch out!' and the next moment something hard hit the toe of my boot and bounced off. There was a blinding explosion, and I staggered back holding my arm over my eyes instinctively. Was I wounded? I looked down rather expecting to see blood pouring out, but there was nothing. I looked for the sergeant who had been beside me. At first I couldn't see him, and then I saw him lying sprawled out on his back groaning. His leg was just a tangled mess. I realized all at once what had happened: one of the enemy in the trench had thrown a grenade at me as he came out with his hands up. It had bounced off my boot as the sergeant shouted his warning, and had exploded beside him. I suddenly felt furious; an absolutely uncontrollable temper surged up inside me. I swore and cursed at the enemy now crouching in the corner of the trench; then I fired at them at point-blank range – one, two, three, and then click! I had forgotten to reload (after firing three rounds at a half-seen target during the earlier stages of the assault). I flung my pistol away in disgust and grabbed a rifle – the sergeant's, I think – and rushed in. I believe two of the enemy were sprawled on the ground at the bottom of the trench. I bayoneted two more and then came out again. I was quite cool now, and started to look around for my pistol, thinking to myself there would be hell to pay with the quartermaster if I couldn't account for it. The firing had died down and groups of men were collecting around me rather vaguely; just then a man shouted and fired a single round. One of the enemy in the trench had heaved himself up and was just going to shoot me in the back when my chap got him first.[7]

The day the bombardment of El Alamein began, troopships nosed out of harbours on the east coast of the USA to bring substantial American ground forces into the war against Germany for the first time, their destination the coast of Morocco and Algeria. In the Mediterranean they were joined by British forces, and on 8 November a combined force of 100,000 men came ashore in Operation Torch, some to a warm welcome, others a hot one, in the face of gunfire from Frenchmen loyal to Marshal Pétain's Vichy government, which had made a separate peace upon the fall of France eighteen months earlier. Elsewhere, too, the tide of war turned that autumn; on the other side of the world, American and Australian troops were beginning to come to terms with the realities of jungle fighting as they slowly drove the Japanese northwards in New Guinea, and far to the north and east of El Alamein, the *Wehrmacht* had reached the limit of its eastwards march on the banks of the River Volga, at the city of Stalingrad, where von Paulus's Sixth Army tried to crush Chuikov's 62nd Army before the onset of winter. But the Red Army would not be the same pushover as it had been in 1941, and its infantry would be particularly tenacious in defence of the city.

Life expectancy for front-line Russian soldiers was measured in days during major offensive operations in the first two years of the Great Patriotic War, as the Soviet high command considered its infantry to be expendable.

Recruitment in the Red Army took place in a haphazard way. In the early years of the war, political commissars would tour large factories and collective farms asking for volunteers, which resulted in whole battalions being formed from individual work places. As the war progressed it became more difficult to generate the manpower required, so more drastic measures were used, with the whole male populations of newly liberated towns and villages being forced into military service. These 'volunteers' were often sent into action without uniforms or weapons as mine-clearing troops, advancing in front of regular units to detonate minefields so the assault could proceed.

The issue of weapons did not follow any set pattern due to shortages. Often soldiers were sent to the front without weapons, and they had to wait until a comrade was killed before they could take his weapon. Training was minimal and was usually carried out at camps near the front or in the trenches. If a Soviet infantryman survived his first battle, he was a veteran and was expected to teach new recruits the ropes. By the end of the war, sections and platoons were organized with nine men forming a section. Each section was supposed to have nine men and two Degtyaryey drum-fed light machine-guns. Although the Russians were keen on sub-machine-guns, these were usually reserved for so-called Guards regiments, with most infantrymen still being issued with the pre-1914 Mosin-Nagant M1944 bolt-action rifle.

Battle tactics at company and battalion level were equally primitive. Junior officers, for example, had little influence over tactics. Battalion attack plans were the norm. Massed artillery, mortar, tank and air bombardments were followed by simple charges straight at the Germans. If the softening-up bombardment worked, then the Russians would be able to mop up the German positions. However, if the German defences survived, then they were usually able to mow down hundreds of Russian infantrymen in the open. Rigid Russian tactics usually meant that the charges would be repeated again and again until units were considered combat non-effective, i.e. down to less than 25 per cent of their original strength. Russian infantry tactics had changed little since the First World War.

In defence, Russian tactical backwardness worked to their advantage. Infantrymen were given a trench or factory building to defend and they would fight on until they were all killed, no matter what the tactical situation. Russian interest in sub-machine-guns, such as the PPD-1940, PPSh-41 and PPS-43, proved invaluable to Soviet infantry defending towns and cities such as Stalingrad. They provided close-range firepower that was crucial for repulsing German attacks, as von Paulus would find out to his cost.

During the month of September, the Sixth Army, reinforced by elements of the Fourth Panzer Army, fought its way through the suburbs of the huge industrial city which crowded the right bank of the river for almost thirty-five kilometres, and into the factories and mills at its heart. Though the Germans' superiority in weapons had meant that virtually all the Soviet armour had already been destroyed, such was the level of resistance that its advance was slowed to a rate of hardly more than a hundred metres a day at best, and frequently not that, with the day's gains often being reversed the following night:

> We would spend the whole day clearing a street, from one end to the other, establish blocks and fire-points at the western end and prepare for another slice of the salami the next day. But at dawn, the Russians would start up firing from their old positions! It took us some time to discover their trick; they had knocked communicating holes through between the garrets and attics and during the night they would run back like rats in the rafters, and set their machine-guns up behind some top-most window or broken chimney . . .[8]

The tanks, which were the Germans' main advantage in the attack, were less useful in the built-up areas of the city proper. Firstly, they could not elevate their main guns beyond about the second storey, and secondly, their comparatively thin rear-deck armour was vulnerable to anti-tank weapons aimed from above. The usual solution was to send *Flammenwerfer* operators along with the tanks, to set fire to any building suspected of holding a nest of defenders, but the job of wielding a flame-thrower was understandably unpopular – the Soviets had a number of particularly gruesome ways of executing those they caught, including crucifixion and hanging from a meat hook under the chin; special rates of pay did little to compensate for that. The only way the Germans were able to keep the job filled, by late September, was by drafting men from the punishment battalions.

The *Wehrmacht* had a significant advantage over the Russian defenders – three to one in men, six to one in armoured vehicles and total air supremacy – and the conditions under which Chuikov's men – and women – lived and fought were extremely difficult, to put it mildly. But if the Germans were to be stopped, there was no alternative, despite the extreme shortage of ammunition and food: the Volga was the last serious barrier before the Caspian to the south and the Ural mountains to the north. It was a 'last ditch' defence, both literally and figuratively:

> ... A soldier would crawl out of an occupied position only when the ground was on fire under him and his clothes were smouldering. During the day the Germans managed to occupy only two blocks.
>
> At the crossroads of Krasnopiterskaya and Komsomolskaya Streets we occupied a three-storey building ... This was a good position from which to fire on all comers and it became our last defence. I ordered all entrances to be barricaded and windows and embrasures to be adapted so that we could fire through them with all our remaining weapons.
>
> At a narrow window of the semi-basement we placed the heavy machine-gun with our emergency supply of ammunition – the last belt of cartridges ... Two groups, six in each, went up to the third floor and the garret. Their job was to tear down walls and prepare lumps of stone and beams to throw at the Germans when they came up close. A place for the seriously wounded was set aside. Our garrison consisted of forty men, but the basement was full of wounded, and only twelve men were still able to fight. There was no water, and all we had left in the way of food was a few pounds of scorched grain.
>
> The infantry attacks stopped, but they kept up the fire from their machine-guns all the time. They attacked again ... and were beaten off. In the sudden silence we could hear the bitter fighting going on for Matveyev Kurgan and in the factory area. Could we

divert even a part of the enemy's forces, and so help the men fighting there?

We decided to raise a red flag over the building, so that the Nazis would not think we had given up. But we had no red material . . . One of the badly wounded men took off his bloody vest, and while the Germans called upon us to surrender, a red flag rose over our heads.

'Bark, you dogs!' cried my orderly, Kozhushko. 'We've still got a long time to live!

We beat off the next attack with stones, firing occasionally and throwing our last grenades. Suddenly, from behind a wall came the grind of a tank's caterpillar tracks. We had no anti-tank grenades. All we had left was one anti-tank rifle with three rounds of ammunition. I handed this to Berdyshev, and sent him through to the back of the building to fire at the tank at point-blank range, but before he could get into position he was captured by German infantry. What Berdyshev told them, I don't know, but I can guess he led them up the garden path, because an hour later they started another attack, exactly where I had placed the heavy machine-gun.

This time, reckoning that we had run out of ammunition, they came out of their shelters, standing up and shouting, advancing down the street in a column. I put the last belt in the gun . . . and sent the whole of the 250 bullets into the yelling, dirty-grey Nazi mob . . . heaps of bodies littered the ground; the Germans still alive ran for cover in panic. An hour later they led our anti-tank rifleman out and killed him before our eyes for having shown them the way to my machine-gun.

There were no more [infantry] attacks. [Instead] an avalanche of shells fell on the building . . . We couldn't even raise our heads. Again we heard the ominous sound of tank tracks. From behind a neighbouring block, stocky German tanks began to emerge. This, clearly, was the end. The guardsmen said goodbye to each other. With a dagger, my orderly scratched on a brick wall: 'Rodimetsev's guardsmen fought and died for their country here . . .'[9]

Slowly, as September became October, the situation changed; for a variety of reasons, some of them political, some logistical, there were no reinforcements forthcoming for the Germans, but the Red Army garrison in Stalingrad received replacement soldiers every night, often ferried across the river and sent straight into battle.

The defenders' tactics, too, had undergone a change, and limited local counter-offensives sometimes took the place of dogged static resistance. General Chuikov himself described how they were to be undertaken, in a rather bizarrely styled communiqué which none the less conveys the tone of this sort of combat:

> Get close to the enemy's positions; move on all fours, making use of craters and ruins; dig your trenches by night and camouflage them by day; make your build-up for the attack stealthily, without any noise ... Two of you get into the house together – you and a grenade; both be lightly dressed – you without a knapsack, the grenade bare; go in grenade first, you after; go through the whole house, again always with a grenade first and you after ...
>
> There is one strict rule now – give yourself elbow room! At every step, danger lurks. No matter – a grenade in every corner of the room, then forward! A burst from your tommy-gun around what's left; a bit further – a grenade, then on again! Another room – another grenade! Rake it with your tommy-gun! And get a move on!
>
> Inside the object of attack the enemy may go over to the counter-attack. Don't be afraid! You have already taken the initiative, it is in your hands. Act more ruthlessly with your grenades, your tommy-gun, your dagger and your spade! Fighting inside a building is always frantic. So always be prepared for the unexpected. Look sharp ...![10]

Almost incredibly, six weeks into the siege the Germans were still making some headway. By the end of October, the Russian perimeter was nowhere more than one and a half kilometres away from the river bank, in an area which was almost entirely filled with heavy industrial plant: foundries, machine shops, assembly shops and the like, an impenetrable jungle of iron and steel. The advance got slower and slower, and the price both sides paid was terrible. 'We have fought during fifteen days for a single house, with mortars, grenades, machine-guns and bayonets,' wrote a young lieutenant of the 24th Panzer Division:

> Already by the third day, fifty-four German corpses are strewn in the cellar, on the landings, on the staircase. The front is a corridor between burnt-out rooms; it is the thin ceiling between two floors.

Help comes from neighbouring houses by fire escapes and chimneys. There is a ceaseless struggle from noon to night. From storey to storey, faces black with sweat, we bombard each other with grenades in the middle of explosions, clouds of dust and smoke, heaps of mortar, floods of blood, fragments of furniture and human beings. Ask any soldier what half an hour of hand-to-hand struggle means in such a fight. And imagine Stalingrad; eighty days and eighty nights of hand-to-hand struggles. The street is no longer measured by metres, but by corpses . . .

Stalingrad is no longer a town. By day it is an enormous cloud of burning, blinding smoke; it is a vast furnace lit by the reflections of the flames. And when night arrives, one of those scorching, howling, bleeding nights, the dogs plunge into the Volga and swim desperately to gain the other bank. The nights of Stalingrad are a terror for them. Animals flee this hell; the hardest stones cannot bear it for long; only men endure.[11]

Desperate fights like this took place all over the city, until the last isolated strongholds in the Krasny Oktyabr, the Barrikady and the Traktor factories, fronting the river to the north, finally fell, one by one, in late October and early November. But by now the Germans were exhausted, and certainly could not summon the energy to cross the river and break out. And then the unthinkable happened: on 19 November, Marshal Zhukov's armies to the north and south of the city suddenly went onto the offensive. By 23 November, the Russians had formed a ring of steel around Stalingrad, and now it was von Paulus who was cut off, along with some 250,000 fighting men, 100 tanks and perhaps as many as 2,000 guns. It took the Red Army two whole months more to crush the invaders who had taken over the city, but by that time, the main fighting front had been pushed back almost as far as the Sea of Azov, 300 kilometres to the west. Hitler's armies were on the run at last.

During the first six months of 1943, the Allied forces in North Africa, advancing from both east and west, crushed the Germans and Italians between them. In July they followed them to Sicily, and mounted Operation Husky, establishing a bridgehead on the south and east coasts of the island on 10 July despite planned airborne landings going badly wrong, thanks to unfavourable weather conditions. Almost incredibly, British and American forces secured the entire island by 17 August, and the Allies had re-established a toe-hold in Europe. One unlooked-for and unexpected result of the invasion was the fall of the Italian dictator, Benito Mussolini, on 23 July. He was arrested, and was later to become the subject of one of the most daring commando raids in history when he was rescued by the celebrated Otto Skorzeny, one of the few Germans to run a successful 'special' operation.

The Allied armies in Sicily wasted no time; in the month after consolidating the island, they launched Operation Baytown, which saw Montgomery's veteran Eighth Army cross the Straits of Messina on 3 September, and Operation Avalanche, which landed the Anglo-American US Fifth Army at Salerno on 16 September, together with other combined operations aimed at Taranto and Bari. The invasion and occupation of southern Italy was completed when, on 1 October, Naples was liberated. By then, the temporary government set up in Rome by Marshal Badoglio had itself fallen, and Italy had signed a separate peace, which meant that the German forces in the huge peninsula were now in law, as well as in fact, an army of occupation. They behaved appropriately; having started easily, the re-occupation of Italy now became a hard, bloody business that was to continue almost to the very end of the war itself.

So little progress was made in breaching the Gustav Line (the defensive position which stretched from north of Naples to Foggia on the Adriatic coast) that on 22 January 1944 a second amphibious landing was made behind it, at Anzio. That, too, bogged down Field Marshal Kesselring's forces containing the Allied troops in an endless series of savage fights around the thirty-kilometre perimeter of the beachhead.

The break-out came on 23 May, following a successful thrust through the Gustav Line by the 2nd Moroccan Division of the French Expeditionary Corps (soldiers who were to continue to excel themselves throughout the battles for Italy), and then a concerted attack in strength involving American, British, French and Polish units. Both the main attack and the break-out from Anzio were badly directed, in strategic terms, by the US General Mark Clark, who only had eyes for liberating Rome; in his short-sighted quest for personal glory, Clark let an entire German army, the Tenth, escape to fight another day, and almost certainly prolonged the struggle for Italy in the process. So determined was Clark that it would be US troops who would liberate Rome that, almost incredibly, he actually gave orders that fighting men of other nationalities were to be held back from the city by force, if necessary.

German soldiers surrender to British Tommies in the First World War. Usually only a few soldiers survived the hand grenades and bayonets used during a violent trench clearing operation.

Heavy losses forced the British to draft in Gurkha troops to help their war effort in France. The Gurkas proved to be determined and brave fighters on the Western Front as here and elsewhere.

Facing page: British Tommies wait to go over the top. By the middle of the Great War trench systems were highly complex affairs with bays to provide overhead protection from shell fire and zig-zag trench lines to prevent shell blasts travelling along the length of the trench.

Left: British Troops rest in a trench on the first day of the Battle of the Somme. This battle cost the British 60,000 casualties on the first day. By the time the Somme campaign ended in November 1916, several hundred thousand troops were lost.

Below: Periscopes were an essential tool of trench warfare in the Great War because they allowed soldiers to observe the front without exposing themselves to enemy sniper fire.

Left: Allied offensives in 1918 broke the deadlock of trench warfare and allowed these US Army soldiers to advance into open country and engage the Germans in fast moving mobile warfare.

Main picture: Bayonet drill has remained an essential skill for the infantryman up to the present. These Second World War Royal Marines are learning how to parry and thrust the weapon.

Below: The rifle butt is also a deadly weapon in close-quarter combat. A powerful butt stroke to the head or other vital organs can kill an opponent.

Top left: British Commandos advance up a street in Normandy to clear out a German strong point, 1944. Sniper fire has pinned them down in a doorway.

Top right: House clearing in Italy. British troops are forcing their way into buildings with rifle butts, before storming in at bayonet point.

Main picture: A British Bren gunner provides covering fire for an advance through a French town. The use of fire and manoeuvre tactics was standard by June 1944 and cut down on the number of casualties considerably.

Below: A German Wehrmacht sniper observes the front line on the Eastern Front in Russia. His comrade is watching the fall of shot with binoculars to improve the sniper's accuracy.

German infantry advance through a Polish street with Panzer and artillery support. These tactics proved unstoppable in the early years of the war.

SS Panzergrenadiers storm a house in Russia, 1942. They are armed with a mix of carbines and sub-machine guns, and carry minimum kit to improve their mobility during fighting in confined buildings.

After the break-out through the Gustav Line and from Anzio, there was little to stop the Allies' march on Rome, and the city was duly liberated on 6 June. It was a momentous enough event – the liberation of the first European capital city from Nazi occupation – but it went almost unnoticed outside Italy in the clamour surrounding the happenings of that same day on the coast of Normandy. We shall return to Operation Overlord and the subsequent battles for France and the Low Countries, but first look at some of the other events the invasion of France overshadowed, particularly those which had taken place in the jungles of Burma, half a world away, over the preceding months.

After their lightning strike on Singapore in the closing weeks of 1941, the Japanese had gone on to occupy all of Indo-China, as well as the thousands of islands south to New Guinea and east across the Pacific.

The Japanese Army was built around its infantry arm. They spearheaded every offensive and were considered almost unstoppable in the early years of the Second World War.

Japanese soldiers were the product of a highly militaristic society. Teenagers were given military training at school and conscripts were subjected to intense brutality during their service to prepare them for combat (which made them liable to commit atrocities against civilian and captured enemy soldiers). They were also brainwashed in the Japanese warrior code from an early age, making them fanatically loyal to their Emperor and afraid of surrender in battle because of the dishonour this would bring on their families back home.

The Imperial Japanese Army was designed for offensive operations in jungle or mountainous terrain. Its soldiers were expected to carry all their own weapons and equipment. Heavy machineguns, mortars and many artillery pieces were designed to be moved by pack horses. In line with the Japanese Army's offensive spirit, its standard light machine-gun, the Type 96, came complete with a 20in-long bayonet, the only such weapon system in the world. The main Japanese infantry rifle, the Meiji 38 Arisaka, was also produced in a variant that featured a folding bayonet. The long bayonet was a Japanese trade mark. It was more than a fighting weapon, it was used to cut through jungle foliage and for the ritual execution of prisoners.

During offensive operations, the Japanese Army had to live off the land or from captured supplies – the use of enemy weapons and ammunition was a standard procedure. When things were going well this was easily accomplished, but if the enemy stood and fought then the Japanese quickly ran out of food and ammunition. Often starvation forced the Japanese to launch suicidal attacks to capture food (surrender or withdrawal were just not considered honourable).

Japanese battle tactics were effective. They not only relied on the Banzai charge, but also used a variety of tactics to win. Patrolling to find weak spots in defences followed by infiltration to outflank and surround enemy positions were common during offensive operations. When resistance needed to be neutralized, the Japanese brought up their pack-horse-carried artillery to destroy enemy strong points with direct fire. This was usually enough to break the will of the enemy, allowing the infantry to sweep in and bayonet all the defenders, even if they surrendered or were wounded.

The Japanese were experts at patrolling and small-unit skills. They struck terror into Allied opponents by their ability to move silently through jungles at night to kill sentries and then charge unsuspecting positions. In defence the Japanese showed little imagination, relying on reinforced bunkers and other field defences. Once the Japanese soldiers were assigned their positions, they were given little choice but to fight to the death. Occasionally counter-attacks were launched to take back ground, but they were usually ill-planned and ended in slaughter.

By the time of El Alamein and Stalingrad, the Japanese had reached as far west as the borders of India itself, and over the course of the next eighteen months were only kept back by a series of desperate actions on the ground and almost constant aerial bombardment. They still pressed forward, none the less, and at 04.00 on 5 April 1944, began what was to become one of the fiercest close-quarter infantry battles of the entire Asian war, when they advanced on the town of Kohima.

By the evening of the first day, the Japanese had the settlement almost surrounded, and the following night, as more fighting units arrived, they launched a series of attacks. They were driven off by the men of the Royal West Kent Regiment (RWKR) augmented by units of the Assam Regiment, and for the next week contented themselves with making preparations for a major assault when the balance of General Sato's 12,000-strong force arrived. On the night of 13/14 April it came . . .

During the Japanese drive into the Arakan region, an incident had occurred which provoked intense hatred in the ordinary British soldier for his Japanese counterpart: the massacre of wounded and medical staff at a main British dressing station. Men lying helpless on stretchers were butchered with sword and bayonet; doctors and orderlies were lined up and shot; Indian bearers were forced to carry back Japanese wounded, and then they, too, were murdered. During the campaign which followed, it became an article of faith not just to repel the attackers, but to kill as many of the invading army as possible, and many welcomed the attack when it came as a renewed chance to put this informal policy into effect. And effective it certainly was: over the following four days, some 7,000 Japanese soldiers were thought to have penetrated British lines, by one means or another; over 5,000 bodies were found, and it is to be assumed that many more died of their wounds in the mountainous jungle.

The RWKR joined the Kohima garrison in its hilltop position on 4 April, just before the first Japanese night attack had overrun some of the British positions. On the morning of 6 April, a platoon of Indian troops escorted 2,300 walking wounded out of the perimeter (a rough square, about a thousand metres to a side), leaving fewer than 3,000 men in all to defend it, some one-third of whom were reckoned as fully fit.

The Japanese began by making attacks on Kohima by day, but the reception was too hot, and they soon changed over to night assaults, though between the infantry attacks, artillery, mortars and machine-guns kept up a relentless onslaught. Because the area was so small, defenders often had the impression they were under sniper fire from the rear, when in fact the incoming fire they were receiving was 'overs' from attacks on other defensive positions. There was no rest, for British and Japanese positions were often within metres of each other, and even the slightest movement invariably brought a shot or a grenade.

Gradually, the British area was compressed under the sheer weight of the Japanese attack, and the defensive square shrank to a quarter its previous size. The whole hilltop, continuously wreathed in smoke, was covered with tall trees, the canopies of which closed out the sky. Consequently, the undergrowth was rather thin. The area was criss-crossed with tracks, running in all directions. Visibility by day was about fifty metres, but by night it was zero. An enemy soldier would suddenly appear at the end of a Tommy's rifle – so reflexes had to be quick.

The Japanese tried many tricks. One would shout, 'Hey, Johnny, let me through, let me through, the Japs are after me. They're going to get me!' The call would be taken up elsewhere around the perimeter. 'Let me through . . .!' But the men of the garrison were no strangers to jungle warfare, and they made no answer. Or they would try irritation rifle fire – single, well-spaced shots, apparently made deliberately at specific targets. Nervous new troops would fire back, the attackers would spot the muzzle flashes and use the information to help them plan their next attack.

At night, Japanese assault troops, wearing soft shoes and with all noise-making equipment removed, sometimes attacked as silently as possible, so that all the defenders would hear was the sound of scuffling. At other times they would reverse the procedure, screaming, yelling and blowing battle bugles. The defenders preferred a noisy attack, for then they had a sound-target, if not a visual one. At such times, in the darkness, self-discipline was of the utmost importance, for no officer or NCO could control more than those immediately next to him. Conditions for the wounded were particularly bad, and some were hit again and again. They were in great pain, and drugs were scarce; they could only 'grin and bear it'.

The stories of heroism were many, but one in particular stands out: on Easter Sunday, a party of Japanese broke into the British lines and established themselves in bamboo huts within the defences themselves, from which they could machine-gun the men of the Royal West Kents. Lance-Corporal Harman of D Company climbed out of his trench and unhurriedly approached the most dangerous machine-gun. At a range of about thirty metres he threw two grenades at the post, silencing it. Then he ran into the hut and shot or bayoneted the survivors, re-emerging with the captured gun. During the course of that same attack, Harman cornered seven Japanese hiding in the disused ovens of the bakery; he dropped grenades in each one.

The following day, the Japanese set up a machine-gun post on a ridge overlooking D Company's position. From there, the Japs could easily dominate the British trenches. Harman realized that if this post were not dealt with immediately, his company would suffer heavy casualties. Ordering his Bren gunner to give him covering fire, he found a place from which he could overlook the machine-gun, and shot dead one member of the gun team. Then he advanced alone, again in an unhurried way, despite Japanese fire now being directed at him. Twenty-seven metres from the machine-gun Harman fired, killing another; then at bayonet-point took the post.

He started back to his own trenches – still moving without haste, despite warning shouts from his own men. A machine-gun burst hit him in the back before he reached safety. 'I got the lot – it was worth it,' he said before he died.

Relief came to the Kohima garrison on 20 April, at 06.00, with the arrival of the 2nd Indian Division, which had taken five days to cover the last three kilometres of enemy-infested jungle from Jotsoma to the west, where they had earlier established a howitzer battery which had been the Kohima garrison's only outside support. The surviving British and Indian troops marched out of the protective enclave, leaving 1,387 of their number behind them.

Fighting in the Kohima region, and at Imphal, to the south, continued to mid-summer, becoming more and more sporadic as the Japanese forces weakened. The Anglo-Indian forces concentrating in the region, including Orde Wingate's celebrated Chindits, then began the slow and painful business of advancing south-east through the Burmese jungle, while the Chinese, together with the remnants of the Americans who made up Merrill's Marauders, and reinforced by a further 2,500 men flown in to hastily constructed airstrips, advanced from the north-east. Meanwhile, the Australian–American campaign in New Guinea was still grinding on, and beginning to bear sweeter fruit (the remaining Japanese, many of them barely able to hold a rifle, so sick were they, were finally expelled from the island at the end of July), while the American assault on the Japanese-held islands further north had finally begun, with landings on Saipan (on 17 June), Tinian and Guam. The resistance there was ferocious, and despite support from the heavy guns of battleships and cruisers, as well as carrier-borne aircraft, the US Marines tasked with retaking the islands had to fight literally every inch of the way against troops who were fully prepared to die rather than give ground, and who never even considered surrender as an option.

One of the marines who faced this fanatical Japanese resistance was Alvin M. Josephy Jr, who fought in the 3rd Marine Division on Guam, during operations in August 1944:

A little before midnight I began to doze. On my back, with my head resting inside my helmet, I was on the slope, half in and half out of the hole. [Corporal Walter] Page was on top of the hole on his stomach, his chin was resting on the dirt; he was gazing

95

straight ahead. Suddenly he began sliding back towards me. He moved his arm back slowly, pinched me, and without turning his head whispered calmly: 'Here they come. Look out.'

I didn't dare turn on my stomach. Arching my head back, I looked upside down at the path. The full moonlight on the little clearing gleamed through the trees and lit up the path. Four figures were moving down in, coming toward us and talking in low voices. Two carried rifles. They were completely unaware of our presence.

I looked across our shell hole. The lieutenant was pressed against the dirt on the opposite slope, breathlessly watching the approaching Japs. Slowly he pushed his pistol out ahead of him. Then his hand twisted it into a firing position. We waited tensely as the Japs came nearer. Finally they were less than ten yards away, about to turn into the ration dump.

A BAR [Browning Automatic Rifle], almost beside us, shattered the night with a frightening roar. Red flames from the muzzle blinded us. The next instant there was a deathly silence, and the blackness returned. A few yards away, we could hear a moaning.

Page looked back wildly.

'One down,' he whispered. 'Where are the other three?'

We lay down as quietly as we could, listening to the wounded Jap moan. We looked around at the black shadows, wondering what had happened to the other three enemy.

Obviously we weren't in a good position. The other Japs, two of whom might have been the ones with the rifles, were probably hiding in the trees, waiting for us to make a motion and reveal our position. The moonlight was streaming down on us in the clearing; while we couldn't see the enemy, it would be easy for them to see us.

We quivered in terror. The wounded Jap's voice began to rise in singsong prayer. It was like a baby's voice crying – high and sad. We could hear the mosquitoes buzzing in our ears and feel them stinging through our dungarees; but we couldn't scratch. A coconut crashed in the jungle, and our hearts beat wildly. Still no sign of the other Japs. The smell of dried fish filled the clearing.

Suddenly we heard a tapping – like metal on a rock.

'Duck!' Page breathed.

The next instant there was an explosion and a blinding flash. Pieces of metal and flesh and chunks of dirt rained down on us.

The singsong prayer was over – the Jap had blown himself up with a hand grenade.

We stayed quiet and on guard the rest of the long night. The mosquitoes tortured us. After a while the dead Jap, lying a few feet from us, began to smell. The odour joined with that of the fish in a stench we could never forget. We peered at the trees, trying to find the other Japs, but they had vanished.

There's nothing like the jungle to make one nervous and tired. The jungle is terror. Stalk and be stalked.[12]

Despite the suicidal defences organized by the Japanese, the American amphibious drive across the central Pacific went on. By April 1945, landings had begun on the island of Okinawa – the last stop before the Japanese Home Islands. Marine Sergeant William Manchester fought on Okinawa, and got pinned down by a Japanese sniper:

[Chet] pulled the pin and threw the grenade – an amazing distance – some forty yards. I darted out as it exploded and rolled over on the deck, into the prone position, the M1 [Garand self-loading rifle] butt tight against my shoulder, the strap taut above my left elbow and my left hand gripped on the front hand guard, just behind the stacking swivel.

My right finger was on the trigger, ready to squeeze. But when I first looked through my sights I saw dim prospects. Then, just as I was training the front sight above and to the left of his rocky refuge, trying unsuccessfully to feel at one with the weapon, the way a professional assassin feels, the air parted overhead with a shredding rustle and a mortar shell exploded in my field of fire. Momentarily I was stunned, but I wasn't hit, and when my wits returned I felt, surprisingly, sharper.

The Jap's slab of rock had my undivided attention. I breathed as little as possible – unlike Chet, who was panting – because I hoped to be holding my breath, for stability, when my target appeared. I felt nothing, not even the soppiness of my uniform. I looked at the boulder and looked at it, and looked at it, thinking about nothing else, seeing only the jagged edge of rock from which he had to make his move.

I had taken a deep breath, let a little of it out, and was absolutely steady when the tip of his helmet appeared, his rifle muzzle just below it. If he thought he could draw fire with as little as that, he must be new on the Marine front. Pressure was building up in my lungs, but I thought I would see more of him soon, and I did; an eye, peering in the direction of my boulder, my last whereabouts. I was in plain view, but lying flat, head-on, providing the lowest possible profile, and his vision was tunnelled to my right. Now I saw a throat, half a face, a second eye – and that was enough. I squeezed off a shot. The M1 still threw a few inches low, but since I had been aiming at his forehead I hit him anyway, in the cheek. I heard his sharp whine of pain. Simultaneously he saw me and shot back, about an inch over my head, as I had expected. He got off one more, lower, denting my helmet. By then, however, I was emptying my magazine into his upper chest. He took one halting step to the right, where I could see all of him. His arms fell and his rifle toppled to the deck. Then his right knee turned in on him like a flamingo and he collapsed.[13]

In all, the United States Marine Corps, which bore the brunt of the fighting in the Pacific theatre, had to take dozens of islands by assault from the sea, each one, it seemed, tougher than the last. Unknown, unplaceable dots on the map such as Iwo Jima became household names for their – and future – generations to display as thrilling proof of a nation's military coming of age. And still there was the spectre of an opposed invasion of the Home Islands themselves hanging over it all, with the likelihood of millions of casualties.

•

Initially, at least (though not right across the front) resistance to the armies committed to Operation Overlord in the huge Baie de la Seine in Normandy was less well organized, though some German units, too, fought to the last man. Desperate to contain the invaders, Hitler rushed elite units of *Waffen-SS* from further south and east in an attempt to create an impenetrable cordon of armour around the bridgehead. His forces were eventually overwhelmed by sheer superiority of numbers and resources, but not before their Tiger tanks, in particular, had wreaked havoc among the lighter, thinner-skinned American and British armour.

The mounting of such an operation against a target prepared to receive it was fraught with risks, as British Prime Minister Winston

Churchill, in particular, well knew. It was he who had, in an earlier incarnation, been largely responsible for the only amphibious operation of the First World War: the landings at Gallipoli. Here, there had been no special preparation; the soldiers went ashore in exactly the same fashion as Marines and Bluejackets had been landed by the Royal Navy in its heyday, a century before:

The leading [Lancashire] Fusiliers' objective was W Beach west of Cape Helles. It was 350 yards wide, with cliffs about 100 feet high at each end and in the centre a mound commanding the beach. HQ and three companies were in position off the coast in HMS *Euryalus* and at 4 a.m. on 25 April 1916 the men transferred to the ship's boats, which were towed shoreward by steam pinnaces.

A belt of wire entanglement stretched across the beach, and the order was passed that the troops would wait behind it until gaps had been made. *Euryalus* and other warships shelled the beach to keep the Turks' heads down. A few hundred yards off-shore the pinnaces cast the tow, leaving the boats to be rowed in by naval ratings. Soon the boats came under heavy and accurate enemy machine-gun and rifle fire.

The troops jumped out of the boats and waded and splashed as far as the wire, then went to ground under the hail of bullets from the front and both flanks. For some time the assault troops were pinned down as most of their rifles were jammed by sand and seawater. A few rifles at last were forced to work and a lucky shot killed a sniper who had caused many casualties.

C Company were the first of those on W Beach to gain some sort of order and start climbing to their objective, Hill 114. D Company, from HMS *Implacable*, had landed at some rocks further left, below Hill 114, and quickly scaling the cliff, surprised and beat back the Turks harassing C Company.

The centre and the right were a bloody shambles. At the wire lay more than 300 men, furiously urged on by officers some distance away. But all those three hundred men were dead or wounded. More and more troops stormed ashore and eventually parties of A and B Companies managed to mount an assault on Hill 138, the objective on the right. They reached the top – and ran into the explosion of a shell from one of the British ships, which mangled several of them.

Soon after 7 a.m., a rough line strong enough to protect the landing-beach from small arms fire had been established but . . . it was nightfall before all the objectives had been consolidated, and by that time, the strength of the 1st Battalion, the Lancashire Fusiliers was reduced to 11 officers and 399 other ranks; 11 officers and 350 ORs were dead or wounded. Major-General Hunter Weston, commanding 29th Division, asked the Fusiliers to recommend six officers and men for the Victoria Cross. According to custom involving acts of 'collective bravery', the War Office reduced the number of awards to three, but after considerable agitation, all six men were eventually awarded their country's premier award, one of them posthumously.[14]

'Six VCs before breakfast', as the Lancashires would proudly say from then until their disbanding in the British Army's 1968 restructuring.

The lessons learned in Operation Torch and in the Italian landings had been valuable, certainly, but chiefly in non-combat areas such as logistics. The real trial had come at Dieppe, on 19 August 1942, when British forces including four Canadian infantry battalions and one of tanks, together with Numbers 3 and 4 Army Commandos and A Commando, Royal Marines, had mounted a raid on the town to see if it was possible to take a French port by assault. It most clearly was not, and the Germans routed the assault force (of the 7,000 officers and men who went ashore, 4,384 were either killed, wounded or taken prisoner) without calling in reinforcements from elsewhere.

In the event, the Overlord landings (strictly speaking, Operation Neptune as far as HWMOST – High Water Mark of Ordinary Spring Tides) did not incur impossible casualties (though certain aspects of the overall operation clearly have to be judged a failure), though a thorough analysis would reveal disproportionate losses in the first units ashore, as one might expect; they were there to 'fill the breach with our Allied dead' to misquote wilfully Shakespeare's *Henry V*. Canadian forces, once again, played a significant role in the British sector (which lay east of the American sector, in the zone east of Arromanche, stretching as far as Ouistreham and taking in Gold, Juno and Sword beaches). A padre with an infantry regiment recalled:

The beach was sprayed from all angles by the enemy machine-guns, and now their mortars and heavy guns began hitting us.

Crawling along the sand, I had just reached a group of three badly wounded men when a shell landed among us, killing the other three outright. As we crawled on we could hear the bullets and shrapnel cutting into the sand around us. A ramp had been placed ahead of us against the sea wall. Over we went. Two stretcher-bearers ahead of me stepped on a mine, and half-dazed, I jumped back down behind the wall again.[15]

Some first-wave units lost more than half their strength on the beaches, in just minutes, cut down by well-placed machine-guns in hardened enclosures. One Canadian company of the 8th Infantry Brigade was thus stricken, on Juno beach, between Courseulles and Bernières, and as the survivors crouched in the cover of the low sea wall, in an attempt to avoid the artillery fire which was adding to the carnage, the driver of a Sherman tank panicked, unable to get his 30-tonne vehicle up off the sand, and drove along the line of men, crushing them. He was only brought to a halt by a Royal Marines officer blowing off a track with a hand grenade. Other atrocities were more deliberate. A young naval rating watched as Canadian infantrymen led six German prisoners into the dunes, and followed, hoping to take one of their helmets as a souvenir. When he caught up with the party, the Canadians had disappeared, leaving the Germans with their throats cut. 'It was the first recorded war crime of the campaign,' wrote one distinguished historian, 'but it would not be the last. Before the invasion of North-West Europe was over, all the Allies – and the enemy, too – would be killing their prisoners in cold blood.'

If the British and Canadians on the left had a bad time of it that morning, the Americans on Omaha beach – 'Bloody Omaha' as it became known, with good reason – suffered even worse, mainly as a result of General Omar Bradley's decision not to use the specially developed armoured vehicles available to them. These modified tanks – particularly the 'Crocodile' flame-throwers, able to send a sheet of flame out to a distance of over a hundred metres, and which we will meet again in due course, as well as those fitted with flails to explode mines and others carrying bridges capable of spanning four-metre-wide ditches proved invaluable, and the American troops who lacked them suffered for that. Omaha was the worst of the beaches; a narrow arc of steep sand dunes between rock outcroppings which had received more than its fair share of defensive preparation thanks to a particularly zealous German sector commander. Nonetheless, sheer weight of men carried it, and by the next morning, the bridgehead there was three kilometres deep; that that was well short of the objective seemed to matter less in the cold light of reality than it had at planning conferences. Nowhere was the objective line actually reached, in any event; the British came closest, between Bayeux and Caen, but even they fell short by 50 per cent. The brunt of the fighting in the lush, green farmland of Normandy fell on the Allied infantry.

•

The British infantry battalions which fought in Europe during the last year of the war were well equipped and trained for modern combat. While there was a high proportion of veterans of the North African and Italian campaigns among the NCOs and senior officers, most of the junior officers and ordinary soldiers had not yet seen action. They were mostly wartime conscripts who had been called up only a year before they were due to go into action in Normandy.

In spite of heavy casualties, the British just about managed to maintain their regimental system, with officers and soldiers being recruited from a specific geographic area and then serving together for the rest of their time in uniform.

The basic infantry unit was the rifle section, which was made up of ten men under a corporal. It was armed with a single Bren light machine-gun, a Thompson or Sten sub-machine-gun and the remainder having bolt-action .303 Lee-Enfield Rifle No.4s. This weapon was not considered to have the accuracy of its First World War predecessor, the SMLE, and the older weapons were highly prized by British snipers.

British infantrymen benefited from the introduction, just before the war, of the 1937-Pattern webbing, which allowed enough food and ammunition for twenty-four hours to be carried in large pouches around a belt supported by a weight-bearing yoke. A haversack could be quickly attached to the yoke straps. To reduce dependency on field kitchens, the British also developed ration packs containing tinned meals and miniature cookers.

Fire and manoeuvre formed the core of British infantry tactics. All movement had to be covered by fire to keep the enemy's head down when the moving part of the section was at its most vulnerable. Terrain was also used to gain an advantage over the enemy, with outflanking movements being used to get behind him or to cover assault teams crawling forward to post grenades in enemy bunkers – there were to be no more Battle of the Somme-style bayonet charges.

When defending, the British infantry platoon built its defence around its Bren guns. Like the Germans, they aimed to cover the platoon frontage with overlapping arcs of fire from each of the section's Bren guns. The Bren was famous for its long-range accuracy out to 600 metres, so section commanders liked to position their machine-guns where they had good fields of fire that could maximize its range. However, being magazine fed, it could not sustain as heavy a rate of fire as belt-fed weapons. Bren gunners had to be very disciplined to conserve their ammunition.

Every infantryman carried entrenching tools to allow the infantry to dig in and prepare defensive positions. Digging-in was the main way the British protected themselves from German mortars and artillery, which were the main killers in north-west Europe. Properly constructed trenches had overhead protection of earth supported by planks or tin sheets to stop shell splinters raining down on the occupants. Only in extreme circumstances would British infantry fight to the last to hold a position. If ammunition ran low or casualties were mounting, then a withdrawal would be ordered, with platoons and sections leapfrogging back, in turn, so troops were always in a position to provide covering fire.

But the battles in Normandy were only the start of what was to be almost a year of non-stop battle as the German occupying forces were driven out of France, the Low Countries and Italy by the Allies, and back through eastern and north-eastern Europe by the Russians. From the outset, they defended stubbornly, clinging on until all hope was gone and then fighting a desperate rearguard action to the next natural line of defence, and as we have seen, such combat inevitably leads both to close-quarter battle, and to a wide dismissal not just of the rules of war, but for even the most cursory regard for the fate of others.

We need not catalogue here the atrocities committed by fighting soldiers during the Second World War – nor even touch on the infinitely greater and more horrible acts perpetrated in the name of politics and racialism – but as they represent a particular aspect of close-quarter battle, we may look quickly at some of the most notorious cases which occurred in the aftermath of the Normandy landings.

Oradour-sur-Glane was a small village near the city of Limoges, in western France, in an area where there was considerable partisan activity in the pre-Overlord period. Part of the invasion strategy was to call up all partisan resources, and around Limoges, the turn-out was high; the nearby town of Tulle, for example, was actually besieged by partisans, and its garrison of rear-echelon troops cut off; forty of them were found murdered and mutilated, and that was afterwards held up as partial reason for the barbarous behaviour of the combat troops who relieved the town, the *Waffen-SS* soldiers of *Das Reich* Division's *Der Führer* regiment, whom we last saw using textbook close-quarter battle tactics to deal with Dutch defensive installations. In the meantime, of course, *Das Reich* had seen extensive service on the Eastern Front, where a more bestial form of warfare was waged; its troops' experiences there, too, have been held up as a reason for their atrocious behaviour in France. *Das Reich* was just one of the elements being sent north to fight the invading Allies in Normandy.

The day following the relief of Tulle, two of the division's officers, Gerlach and Kampfe, were abducted by partisans. Kampfe was shot and killed, but Gerlach managed to escape. On rejoining his unit he said that he had been taken, at one point during his brief captivity, to a village he identified as Oradour-sur-Glane. SS-*Sturmbannführer* Diekmann was sent to investigate, with a company of troops. He sealed the village off, rounded up the inhabitants and herded them into two barns and the church, where he ordered them all executed, by shooting, hand grenades and fire. Six hundred and forty-two people of all ages and sexes, including 207 children, died; just one escaped (by jumping through a small window in the church). Oradour-sur-Glane remains today as it was that day (though a modern village has sprung up close by), as a monument to its dead. Diekmann was later killed in action, and the court martial proceedings against him, instituted by his commanding officer, were dropped. There have been other reasons advanced for Diekmann's behaviour, besides his fury at the way German soldiers had been murdered; the most fanciful concerns an apocryphal horde of gold which the regiment was said to have acquired in Russia, and which, according to a completely unsubstantiated story, was stolen by the partisans who kidnapped Gerlach and Kampfe. Failing to find it, Diekmann laid waste to the village where it had supposedly been taken. It is also suggested that Oradour-sur-Glane was the wrong village anyway; there is another Oradour, Oradour-sur-

Vayres, twenty kilometres away, which was said to have been a partisan stronghold. It was there, perhaps, and not to Oradour-sur-Glane, that Gerlach was taken. The truth is unlikely ever to be known, and in any event, has little bearing on the actions of Diekmann and his men, even if it would go some way towards clarifying their motives.

Six months after this slaughter of civilians, another massacre took place, this time at Malmédy in Belgium. The reasons for this atrocity, however, are less mysterious. A group of US prisoners was assembled at a crossroads near the small Belgian town on 18 December, guarded by two tanks. For no apparent reason, a crewman in one of the tanks, a Romanian of German origin named Georg Fleps, fired his pistol into the group, whereupon the terrified men broke and scattered. Technically, at least, now being guilty of attempting to escape, they were shot down by the tanks' machine-guns and any injured finished off with pistols at close range. The only unknown in this case is the precise number of US servicemen involved: German records, such as they are, say something of the order of 20; Belgian eyewitnesses say around 35, and the US Army maintains that the death toll stood at over 120. After the war, over 500 *Waffen-SS* men, including *SS-Oberstgruppenführer* Sepp Dietrich, *SS-Gruppenführer* Hermann Priess and *SS-Obersturmbannführer* Joachim Peiper, were arrested and charged with complicity in the crime, 42 being sentenced to death and 28 more to life imprisonment. It soon became clear that the judgements had been obtained with tainted evidence – US military policemen had used physical and psychological torture to extract 'confessions'. The death sentences were commuted and Dietrich and Priess were later released, but not until 1955 and 1956, respectively. Dietrich, who was certainly not present, or even anywhere near Malmédy, when the massacre took place was arrested on suspicion of having issued a 'take no prisoners' order, though no concrete evidence of one has ever come to light, while Priess and Peiper were charged with transmitting that order. In the aftermath, however, US Army elements did issue at least two Regimental Orders (which survive) forbidding the taking of German prisoners, particularly if they came from the SS ('No SS troops ... will be taken prisoner, but will be shot on sight,' reads one, issued by the 328th Regiment). The overall effect of the massacre was to introduce a new element of officially condoned – and accepted – brutality and bitterness into the war in western Europe, bringing it more into line with the established practice on the Eastern Front and in Asia.

•

The soldiers of the *Waffen-SS* had learned their brutality on the Eastern Front, but they had also become excellent soldiers. By the fourth year of the Second World War, the *Waffen-SS* panzer divisions had become the elite shock troops of the Third Reich. They were heavily armed and trained for high-intensity mobile warfare, to allow them to spearhead attacks or become the anchor of any defensive line.

The core of each SS panzer division were its two *panzergrenadier* (mechanized infantry) regiments. They followed up behind the armoured fist and cleared out enemy infantry, before going firm and fanatically defending ground.

An SS *panzergrenadier* section contained nine men commanded by a Sergeant (*Scharführer*) as they rode into battle in a SdKfz 251 armoured personnel carrier. The section was almost all armed with automatic weapons, including three belt-fed MG42 general purpose machine-guns, three Schmeisser MP40 sub-machine-guns or the new MP43 assault rifles, and the remaining three men Mauser Gewehr 1898 bolt-action rifles. Every SS soldier also carried a pistol and plenty of stick grenades. The SdKfz 251 was also able to carry large amounts of ammunition to allow the section to maintain an impressive rate of fire for long periods of time. The MG42 was the best machine-gun of the war, with a ferocious rate of fire of some 1,200 rounds a minute.

Hitler lavished his prized SS units with equipment. Only the best was good enough for them. They were one of the first military units ever to be supplied with camouflage clothing, and the stylish SS splinter camouflage smocks became one of the organization's hallmarks during the war.

Not surprisingly, the SS used this firepower to excellent effect in both defence and attack. When advancing, the MG42s were used to lay down heavy suppressive fire on enemy positions to allow the riflemen to advance to close range and post grenades in enemy trenches. If the APC could be brought up, its own machine-guns added to the weight of fire. Known to allied troops as 'Spandaus', the German machine-guns were universally feared because of the way they dominated the battlefield.

In defence the SS *panzergrenadiers* were even more deadly. The machine-guns again played a key part in their tactics, with all the guns in a section being positioned to give them fields of fire covering the frontage of the section positions. This meant that even if one or two guns were knocked out or had stoppages, fire could still be rained down on any enemy advancing towards the position. This was particularly important in Russia, where Soviet human-wave charges could only be stopped with overwhelming firepower. A wall of lead alone kept the Russians out of the German trenches, so if anything went wrong with a section's machine-guns the SS were in trouble.

SS platoons and companies layered out their defensive positions in depth, so if one section was driven back the other positions could bring fire to bear on the old trenches to support counter-attacks. Again, the high numbers of machine-guns in SS *panzergrenadier* sections made it costly for allied infantry to try to press home their attacks against *Waffen-SS* positions.

•

Senior US officers, including General George Patton, welcomed the opportunity to use SS massacres of Allied POWs to stiffen their men's resolve. President Roosevelt himself was not immune. On learning of the massacre, he commented to his Secretary of War: 'Well, it will only serve to make our troops feel toward the Germans as they have already learned to feel toward the Japs.'

The reader may recall, at this point, the killing of five Germans on Sword beach, and the comment that from then on, men on both sides routinely killed prisoners, and wonder how the massacre at Malmédy was in any way different. It is difficult (actually, impossible) to know if Fleps's action was premeditated, but that clearly does not alter the fact that he committed cold-blooded murder; one may also argue that any individual soldier who made a reasoned decision to kill an enemy who wished to surrender, rather than taking him prisoner, with all the inconvenience and possible danger which that might incur, also committed cold-blooded murder. The deciding factor between what constitutes legitimate and illegitimate killing in wartime must surely be tactical necessity, but who decides? The aggressor, in the heat of the moment? Or those who come after, and cannot possibly know the full facts of any case? The question arises in combat of all sorts, but none provokes it more clearly than does close-quarter battle.

Such acts as took place at Malmédy were by no means confined to one side, and the US Army's 45th Infantry Division had a particularly bad record in this respect; in the summer of 1943, for instance, war correspondent Alexander Clifford saw a group of its men fire into a truck full of Italian prisoners, killing all but a few, and later saw others kill sixty more prisoners; while in Sicily, two notable incidents took place on the very same day: Sergeant Barry West, who had been ordered to escort thirty-six prisoners to the rear, shot every one of them at the roadside, while Captain Jerry Compton lined forty-three German snipers up against a barn and machine-gunned them. General Patton's comment on hearing of that last incident was 'Tell the officer to certify the men were snipers or had attempted to escape, or something.' (Both West and Compton faced courts martial and were convicted, but both were subsequently returned to their units; both were later killed in action.)

The battle that *Das Reich* Division was on its way north to fight was to take place in the 'Bocage', the narrow lanes and small fields of Normandy. It was here that German technology showed its superiority, the Tiger tanks in particular proving vastly superior in both armour and armament to their British and American counterparts. Twenty-five-year-old Leutnant Michael Wittmann commanded a company of such tanks in the 12th SS Division. He had already won the Laurel Leaves to his Knight's Cross for having destroyed no fewer than 119 Russian tanks on the Eastern Front. Separated from his command, Wittmann suddenly came face to face with a column of British half-tracked infantry carriers under the protection of Cromwell tanks. After some initial hesitation, Wittmann engaged the infantry vehicles, sending them up in flames one after the other, and when he was confronted with a Cromwell, its 75mm solid shot bounced off the Tiger's armour while his own 88mm gun put the British tank out of action with a single round. A dozen more Cromwells joined the fight, while four more Tigers appeared at Wittmann's side; within minutes the British tanks were all either destroyed or desperately running for cover. Wittmann's tank was eventually put out of action too, when a round from a six-pounder anti-tank gun took off one of its tracks, but by that time the damage was done.

But the superiority of German armour apart, the nature of the Bocage countryside, laced with thick hedges, many of them the best part of half a millennium old, was the defenders' main asset, and proved a most effective killing ground. The US 2nd Infantry Division, for example, tried to take Hill 192, 6 kilometres to the east of St Lô, in early June, and lost 1,200 men in three days; a month later, they were still trying to take it. The 29th and 35th Divisions fared little better, and in all, the twelve US Divisions committed to the battle lost 40,000 men over a seventeen-day period, and advanced just twelve kilometres. Much of the fighting was desperate:

A man staggered up to the command post and nearly fell into Martin's arms. Only some instinct had prevented the CP guards from shooting him, as he had made no attempt to answer their low challenge or to halt. It was an officer, a forward observer who had been with Able Company [A Company, 1st Battalion, 115th Infantry Regiment; part of the US 29th Infantry Division, which had landed on Omaha beach and didn't spend a single day out of combat afterwards until the liberation of Paris. Able Company's command post was overrun before St Lô].

He was panting, incoherent. Martin and Johns half carried him down the steps into the command post, where they sat him on an ammunition box that served as a chair. In the light of the single candle he made a horrible sight. Blood was smeared over his face, his hands and his shirt. But Martin could find no wound. The man was dazed, dumb. He sat with his head down, still panting, paying no attention to questions. He stared at the blade of a hunting knife in his right hand. It was still sticky with blood as he raised and lowered it. He seemed almost to be weighing the knife. He opened and clenched his fingers, never taking his eyes from the red-stained blade.

Martin heated some coffee and forced it gently on the man, who was a close friend. The young officer took the coffee and began to settle down. The wild look went out of his eyes. His trembling lessened until it almost ceased. Finally, he began to talk, the words falling out in chunks, sometimes incoherent, 'We were out on Able OP2. I was asleep. The stuff started falling all around us. Before I knew what was going on, the Krauts were in there with us. They killed two Able boys – right in front of me. I held up my hands and my radioman did too. They took my pistol belt. Then the others went on, and one Kraut pushed us towards the rear. His rear. We walked a long time . . . I don't know how far, or where we were. Then some of our own mortars came in

109

close. We stopped, and the Kraut looked around for cover. There wasn't any, so we ran a ways, then the artillery started. It was hitting in front of us. We stopped again and found some old German foxholes. They had tops on. The guard turned his back on me. He was kicking at a hole to see if he could get in it, and my radioman jumped him. My man jerked off his helmet, hit the Kraut in the back of the neck with it. Then I remembered that I still had my knife.' He pushed the blade forward and looked at it for a long time, his fingers still closing and opening around the hilt.

'I jerked it out.' He stopped talking and put his head down almost to his knees, still holding the knife at an awkward angle in front of him. He didn't move for a full minute, then, with an effort, he sat up again and went on. 'I jumped him just as he turned round. I hit him with the knife and it went into him. Then I hit him again and again and again. I couldn't stop hitting him even after I knew he was dead. My radioman pulled me off him, finally, and we started back.'

Major Johns was bouncing with impatience to ask questions but something held him in restraint until the man stopped. 'Where were you? How did you come back? Did you see any more Krauts?'

The lieutenant looked at him dully, as if he didn't understand . . . It was obvious they were going to get no useful information out of this man. He was another combat exhaustion case, shocked more by the awful experience of killing a man with his own hands than by the fire he had been under or the fact that he had been captured. The major motioned to Martin, who led the still-dazed man away to the aid station.[16]

To try to break the deadlock in the Bocage, over 2,000 American aircraft were tasked to bombard enemy concentrations in an area seven kilometres wide by five deep with high explosives and a new weapon, napalm, but missed their targets completely – twice – causing considerable numbers of American casualties (over 100 killed outright, with 700-plus injured) instead.

Further east, the British and Canadians were involved in Operation Goodwood, to take the city of Caen, which stood in the way of them beginning a pincer movement against the Germans defending the Bocage. They battered it night and day for a month, in fighting very reminiscent of that which had reduced Stalingrad, and when it finally fell (followed a week later by St Lô, the other

German bastion in the region, of which one GI, looking at the ruined city, said: 'We sure liberated the hell out of this place'), the floodgates opened, and tens of thousands of German troops surrendered and almost as many were killed, surrounded in vast pockets, cut off from any hope of resupply. None the less, many escaped through what became known as the Falaise Gap, and it was not until late August that the Allies finally broke out of Normandy proper, to the south and east, crossing the Seine and securing bridgeheads there, and then liberating Paris, on 25 August. Meanwhile, the Americans and French had also landed on the Riviera, between Toulon and Nice, and had begun to close a much more massive pincer designed to sever the entire western half of France. They made good progress, and by the end of the month were threatening Lyons, just 500 kilometres from the French capital. And by then, the Red Army had driven the Germans back in the east, too – almost back to the line from which they had jumped off to begin Operation Barbarossa, three long years before.

The long retreat from the high-water mark the Germans had reached in the autumn of 1942 was hard fought all the way, and much of the combat was desperate, particularly as the command structure in the German Army started to fall apart under the pressure:

During the night of 6 January 1944, the Soviets launched a surprise attack on Hill 159.9 [on the Ukrainian Front, near Kirovograd], dispensing with the usual tank and artillery support. The Soviet troops, masters of stealth, crept up the hill and into the trenches. Lepkowski was awakened from a restless sleep by a piercing scream.

From outside the command post a paratrooper shouted the alarm. The Leutnant reached for his sub-machine-gun. The men of the company headquarters squad likewise grabbed their weapons. They ran out of the command post – right into the enemy troops.

'The Russians are in the trench!' The main line of resistance had been lost, but Lepkowski acted without hesitation. He had to drive out the Russians before they solidified their hold on the hill.

'Let's go! Counter-attack!' They were only six men, but they were all armed with machine-guns, and they cleared the trenches with a murderous hail of fire.

Lepkowski's men [of the 5th Company, 2nd Regiment, 2nd

Fallschirmjäger – Parachute – Division] were driven off the hill next day, and fought a hard-pressed rearguard action while falling back on the single remaining bridge over the River Ingul; their resistance bought enough time for the [SS-] *Grossdeutschland* and *Totenkopf* Divisions to reinforce the sector, however, and the retreat did not turn into the rout which had earlier seemed unavoidable.

The Ingul bridge had been blown by the time the remnants of the 5th Company got to Kirovograd, just a few remaining girders spanning the gap . . . 'Well, that's it, the bridge has gone,' said a dejected paratrooper. Trying to swim the ice-cold river would have been tantamount to suicide. Under cover of the darkness, they crept closer.

'We can get across!' said Lepkowski confidently. They crawled slowly and cautiously on to the approach to the bridge, which was still standing, and then over projecting crossbeams. There were large gaps to cross. But then the Russians spotted movement and a machine-gun began to fire. A second and a third joined in and then the night was shattered as several tanks and anti-tank guns opened up. Lepkowski urged his men on. There was a scream as one paratrooper was hit and fell into the river. Filled with a powerless rage, Lepkowski fired off an entire magazine in the direction of the enemy. A machine-gunner moved his weapon into position and opened fire on the muzzle flashes on the Russian side. Finally, under cover of the machine-gun, the survivors reached the west bank.[17]

Despite their heavy losses, the paratroopers were sent into battle again with very little rest, in a southern sector in a huge loop of the River Dnestr, where they took part in an energetic counter-attack which saw 10,000 Russian soldiers taken prisoner after what amounted to a foot race between victor and vanquished. But then, in mid-March, the division was pulled out and sent back to Germany for rest and to train replacements; the next time they saw action was at Brest, on the Atlantic coast of France, on 22 June.

Back in France, the break-out from Normandy, once it finally began in August 1944, pressed on with speed. Once across the Seine, the British moved north into countryside which had regularly been stained with their blood over the previous centuries, through Picardy and into Flanders and the Low Countries, with the double aim of neutralizing the V-weapon sites which were starting to rain unmanned flying bombs and rockets on London, and of opening up another major port, Antwerp, prior to the assault on Germany itself. Meanwhile the Americans, on a wider front, headed across the open plains of Champagne and into the hills of Aragonne and the Vosges, where an earlier generation of 'Doughboys' had fought in 1918, towards the old Franco-German border in the Rhineland. Further north, in the forest of the Ardennes, the following Christmas, they were to get what was probably the biggest shock of their collective lives when the Germans counter-attacked in force.

The British and Canadian troops met little initial resistance, and entered Brussels on 3 September and Antwerp on the following day, cutting off the German Fifteenth Army in the process, and capturing or cutting to pieces most of the Seventh Army, which had escaped from the Falaise Gap. A turn to the east at this point would have taken Montgomery into Germany's industrial heartland, the Ruhr valley, but internal pressure on Eisenhower, the Allied Supreme Commander, saw to it that Montgomery's troops were first tasked with driving into Holland, while Patton's Third US Army, allocated the lion's share of the available supplies, was given the go-ahead to cross the Moselle river and force the Siegfried Line, Germany's border defences, and be first to enter the Reich itself. As part of the plan to outflank the Siegfried Line to the north, known as Operation Market Garden, Montgomery's 1st Airborne Division, together with the American 82nd and 101st Airborne Divisions, were to undertake the biggest airborne operation ever. The object was to seize a corridor eighty kilometres long from Eindhoven to Arnhem, straddling three major rivers – the Maas, the Waal and the Rhine. Along this causeway British and Canadian armoured columns would then advance. It proved to be a failure, largely due to being overambitious. (The same judgement should be made of the German offensive in the Ardennes, three months later.) None the less, it has since gone down in history as an example of heroism *en masse*. For the 'Red Devils' captured and held what Lieutenant-General F.A.M. 'Boy' Browning had described from the beginning as 'a bridge too far' – the Rhine crossing at Arnhem – for nine days, with very little but small arms, before, their supplies exhausted, they were forced to surrender or die. Of the 10,000 British troops committed to this phase of Market Garden, 2,163 escaped; approximately 1,130 died in the battle, and 6,450 were marched into captivity. Browning meant that the offensive was too far-reaching in its objectives, and should have been limited to seizing and holding the Maas and Waal crossings; he was right, but the plan called for a following offensive into the Ruhr by the back door, so to speak, from north of the Rhine.

From the very start, it had been clear that the Allies would need a very large helping of luck if the operation was to be successful, but in the event, good fortune was distinctly absent (that the 9th and 10th SS Panzer Divisions should both have been resting and refitting in the Arnhem area was certainly a matter of gross misfortune). The situation was compounded, too, by what was seen as the poor performance of an elite British Division, the Guards Armoured, which was intended to spearhead the drive to Arnhem and beyond. As the head of the US 504th Regimental Combat Team, Colonel Tucker, put it:

> We had killed ourselves to grab the north end of the bridge. We just stood there, seething, as the British settled in for the night, failing to take advantage of the situation. We couldn't understand it. It simply wasn't the way we did things . . . especially if it had been our guys hanging on by their fingernails, eleven miles away. We'd have been going, rolling without stop. That's what George Patton would have done whether it was daylight or dark . . . They were fighting the war by the book. They had 'harboured' for the night.[18]

Tucker's opinion was widely shared – though it was seriously flawed; the tanks in question were not capable of fighting at night, and had considerable difficulty even moving after dark, except along well-defined roads; along the route in question they would have been hopelessly vulnerable. But as news of the British armour's failure to press on spread, a coldness developed between the two major allies in Western Europe which, many historians believe, spelled the end of effective military co-operation on the Western Front. Meanwhile at Arnhem itself, where Colonel John Frost's 2nd Battalion, Parachute Regiment, had indeed gained control of the road bridge across the Rhine, the men of the 1st Airborne Division's fingernails were starting to get very tired, and by this time heroic deeds did nothing to bolster morale, but were seen, instead, as signs of a dread necessity. There were many such, as when Captain J.E. Queripel, from the only area of the United Kingdom the Germans ever occupied, the Channel Islands, ordered his men to make a run for it from an impossible situation and then turned to face the oncoming enemy armed only with the unit's remaining grenades and one magazine in his Sten. Queripel was never seen alive again, and was one of five men to win the Victoria Cross at Arnhem, only one of whom actually survived to receive it.

For the men of the 1st Airborne Division, who had seen action in North Africa, Sicily and Italy, the battle of Arnhem was their first taste of fighting in a built-up area, house to house and street to street. They proved so good at it that an SS major who took Major Freddie Gough, by then 2nd Battalion's senior surviving officer, prisoner, saluted him and said, 'I wish to congratulate you and your men. You are gallant soldiers. I fought at Stalingrad, and it is obvious that you British have much experience of street-fighting.' 'No,' said Gough, 'this is our first effort. We'll be much better next time.'

Peter Stainforth, a young lieutenant in the Royal Engineers, serving with the 1st Airborne Division, was one of the comparatively lucky ones, in that though wounded and captured, he survived. He was to learn that urban combat doesn't necessarily mean just streets and houses, but that no matter in what circumstances one finds oneself, quick reactions are what count most at the very short range one is likely to encounter an enemy. Cut off from his own 1st Parachute Squadron near the centre of the town, and trying to link up with either the 1st or the 3rd Battalion, he found himself in Den Brink Park with his four-man party:

This park had sometime been a flak site for a heavy battery. The great square sandbagged pits were still there, and the undergrowth had been allowed to advance unchecked. Rhododendron bushes straggled in dark-green islands among tall copper beech trees, and bracken and smaller bushes had grown up thickly, reducing visibility to fifty yards at most.

We spread out and moved into the wood rapidly but carefully, with weapons cocked. We had not gone more than twenty yards when things happened. Someone fired at us from a few yards' range with a Schmeisser machine carbine, then leapt up and ran. His burst missed all of us, and instinctively I fired a long burst at the green figure before he dived into a large patch of rhododendron. My shot hit him; the man gasped but plunged on, making rasping noises between his teeth. My blood was up, and without further thought, I gave chase. I imagined that the German was a lone sniper hiding up in the wood. I knew that I had winged him, but he still had a Schmeisser slung round his neck and we could not go on in safety with him at large. Chepstowe [Stainforth's bodyguard and batman] and I ran to the right of the rhododendron clump while I shouted to the others to go round to the left.

Of course, it all happened in a flash. The German and I fired almost simultaneously, and then I was sprinting round the clump to catch him as he reappeared. I ran on for about a hundred yards, across a road and into a clearing, then dropped down on one knee to see where he had gone. I saw him lying behind a tree and two other men bending over him. These two were wearing mottled camouflage smocks and scrimmed helmets and looked not unlike British paratroopers. I peered for a second undecided, but one of them raised his rifle and fired in our direction. He was wearing black leather equipment and as he turned his head, I could see the shape of his helmet. I shouted, 'Hande Hoch!' but when he swung his rifle towards me I gave them a couple of eight-round bursts. They rolled over without making a sound and lay huddled together.

I have only the vaguest impression of what followed. I saw the whole wood running with figures in green and mottled clothing, men jumping from half-dug trenches and throwing away picks and shovels, men scrambling for rifles propped against trees: the place was alive with Germans dodging and scurrying for the shelter of solid trunks. It was too late to run for cover myself, so I just hosed with the Sten at anything and everything that moved. There was no time or need to aim; they were in a semicircle from twenty to thirty yards away. Two men with a machine-gun, a Spandau, tried to swing it around at me but died over their gun before they ever fired a burst. Two others fell as they dived for trees, another looked out from behind his cover and then crumpled up. I saw the dust fly up in the faces of two others peering over a bank twenty yards to my right, and they disappeared backwards; a single boot came up over the top and stayed there.

I kept thinking to myself: 'Why haven't I been killed yet? Another second, perhaps . . . why on earth am I still alive?' A long splinter of wood flew from the tree to my right; a twig kicked up by my knee; the bullets were cracking very near. The Germans were rattled and were loosing off anywhere in my direction. As long as my own miniature blizzard lasted I stood a chance; when my ammunition had gone . . .! I clipped on the fifth and last magazine and almost gave up hope. Then I heard my own Bren thundering away quite close, so I leapt to my feet and raced for cover, spraying the Sten in an arc behind me. Clarkson finished off his second magazine and then ran too as I came level with him. Of course he saved my life.

We made that thirty yards in record time. We went over the road and towards a red brick house, then dived for a clump of thickly planted trees. I had two yards to go, then felt the bullet smack into my side. I did the last couple of steps and flopped down, gasping.[19]

The worst of the fighting at Arnhem came around the four buildings which dominated the north side of the huge bridge, which were all that Frost had the men to hold. There were just not enough to form an all-round defensive perimeter, and as a result, the Germans were able to infiltrate the position with ease, though at considerable cost. The building that the small detachment of sappers with the party was detailed to defend – wholly inadequate in number as they were – was a school, and it was vulnerable to attack from the neighbouring buildings:

At midday the fighting flared up again, and this time the trouble came from the rows of buildings behind the school. From these, a machine-gun was now firing in through the staircase windows, preventing communication between the floors; and under cover of this the Germans put in another attack on the crossroads to the south-east. Once again the Bren gunners in the southern wing joined in enthusiastically, while others set about wiping out the machine-gun opposite, using the same ruse with a decoy Bren as before [firing the LMG by 'remote control' – i.e., a piece of string tied around the trigger! – while two others were held in readiness further down the building].

Soon after dark, the Germans made their first attempt to burn the sappers out of the school buildings. It started with a hail of rifle grenades from the neighbouring houses into all the windows on the north face. These small bombs caused enough confusion to allow a German flame-thrower to go into action, and this set fire to the half-tracks resting up against the two wings of the school. As these blazed away, the walls became very hot, floorboards began to char and sparks threatened to set the wooden roof alight. Anyone moving outside the building was silhouetted against the flames and came under heavy fire, but eventually two brave men crawled out with large explosive charges and blew both vehicles to pieces. The concussion shook the school to its foundations, but it did the trick.

The enemy now fired the office building immediately to the north, and the strong wind blew showers of burning fragments over the school. All available men [there were about fifty in the defensive party altogether] were rushed up to the attic with sand, shovels and extinguishers, and a long battle against the flames ensued. The whole area was bright as day and these parties came under constant fire, though covering fire was provided as far as possible from below. At last, after three hours, the flames were under control.

In the small hours of Tuesday morning when the fire next door had also died down, the enemy launched a new attack, preceded as before by a hail of rifle grenades through the north-facing windows. As our men crouched down the Germans swarmed over the wall from the office garden. Then they wilted before the Brens and finally broke and withdrew as they were met with grenades.

Suddenly there was a numbing crash in the southern wing. The passage was filled with clouds of rubble-dust and those awful moans which tell of badly wounded men. The south-west room had received a *panzerfaust* – a 20-pound anti-tank bomb. Part of the wall came down, exposing all the floors, the roof sagged in a tangle of smashed timbers and boards, and rubble lay in heaps. All those in the room were wounded, and the rest were badly shaken and dazed.

When 'Paul Mason' [the sapper officer in charge; Stainforth changed his name in his book *Wings of the Wind*, as he did those of all the other men in his unit, 'for the sake of relatives'] had recovered sufficiently to have a look round, he was amazed to discern groups of the enemy collecting in the shadow of the trees and bushes outside, thinking they could not be seen. A machine-gun team was setting up its weapon in the garden, to fire at the buildings to the south. Mortarmen were carrying up their pieces and ammunition. All round the school, German infantry were standing about unconcernedly in clusters. There must have been two platoons of them.

Quickly our men were warned. Cautiously they crept up to the windows – new magazines clipped into automatics, pins pulled out of grenades. On the signal, everything was let loose at the Germans below. The men stood up in the windows like avenging furies, flinging their grenades and spraying death, shouting in triumph as the grey crowd melted. Everywhere bodies lay huddled together in knots, others doubled up or splayed out where they had been struck down . . .[20]

It was no use, of course. Cut off from resupply and reinforcement, the paratroopers of the 1st Airborne Division had but two choices – fight to the death, or surrender. That they chose the latter is not surprising, but one Arnhem veteran summed up their spirit many decades later when he met the question: 'Would you have done it again?' with something approaching contempt. 'Christ,' he said, 'of course we would have done it again!' Even in captivity, the men showed their true spirit. The walking wounded were taken to the hospital of St Elizabeth, in Arnhem. Peter Stainforth, who was there, said:

I felt a shiver of excitement run down my spine. I have never been so proud. They came in, and the rest of us were horror-stricken. Every man had a week's growth of beard. Their battledress was torn and stained, and filthy, blood-soaked bandages poked out from all of them. The most compelling thing was their eyes – red-rimmed, deep-sunk, peering out from drawn, mud-caked faces made haggard by lack of sleep, and yet they walked in undefeated. They looked fierce enough to take over the place right then and there.[21]

Their unwounded colleagues were even more impressive, at least to the PoWs they joined in the cage at Fallingbostel. 'They were marched along the road past our camp by Regimental Sergeant-Major Lord,' wrote Sergeant John Dominy, a British airman who had spent the last four years 'in the bag'. 'He had the swagger of a guardsman on parade. Their guards, a shambling, dishevelled lot, were just about keeping pace with the steady, Praetorian tread of the finest soldiers in the world. We came instinctively to attention, and Bill Lord, noticing our two medical officers standing with us, gave his party "Eyes right!" and snapped them a salute which would have not been out of place at Pirbright or Caterham.'[22]

On 8 November, Patton launched his assault on the Rhineland, throwing his troops over the Moselle despite some of the worst weather in living memory having swollen it and every other river to the dimensions of great flowing lakes. Further north, Bradley's army made more cautious progress through the wooded hills of the Ardennes, until by the month's end, the Allied front line stood almost north–south, from Nijmegen in Holland to the Swiss border at Basle.

As early as the preceding August, Hitler had ordered that a force of twenty-five divisions must be ready to launch a counter-attack during the month of December, and much to everyone's surprise (not least the Allied High Command's) three new German armies had indeed come into being, scraped up from youngsters of fifteen and sixteen, the scourings of prisons, civil servants and other rear echelon workers who had probably never contemplated going to war, even in their worst dreams. During the first two weeks of the month, they were assembled, under an exemplary cloak of secrecy, along an eighty-kilometre strip of the front between Monschau and Echternach, where they were bolstered by the Fifth and Sixth Panzer Armies under Dietrich and von Mannteufel to the north and the Seventh Army under Brandenburger to the south. Opposing them directly were two US Corps, the VII and VIII, part of Bradley's First US Army. Altogether, over 200,000 German soldiers were to carry out Operation Wacht am Rhein, equipped with more tanks and guns than any German army had seen since Normandy, spearheaded by paratroopers dressed in American uniforms who were to be dropped well ahead of the advance to spread confusion.

From the beginning, Sepp Dietrich, for one, had grave misgivings about the operation's feasibility. 'All I had to do,' he said, 'was to cross the river [the Our, which flows south into the Moselle], capture Brussels and then go on to take the port of Antwerp. The snow was waist deep, and there wasn't room to deploy four tanks abreast, let alone six armoured divisions. It didn't get light until eight, and was dark again by four, and my tanks can't fight at night. And all this at Christmas time.' Field Marshal Gerd von Rundstedt was equally dismissive. 'Antwerp?' he asked, rhetorically, 'If we reach the Maas we should go down on our knees and thank God!'

In the event, the attack, launched at 05.30 on Saturday, 16 December, was contained in the north, save for where the I and II SS Panzer Corps broke through, between Malmédy and St Vith. The main thrust came to be centred on the crossroads town of Bastogne, which was surrounded by 20 December (its defensive garrison was made up largely of men from the 101st Airborne Division, hastily recalled from leave they had been granted after the success of their part in Operation Market Garden). In that southern sector, entire bodies of American troops, many of them inexperienced (the 106th Division, for example, had been in the line for just five days), simply turned and ran before the German onslaught, though many were captured where they stood. Few held out and tried to make an orderly withdrawal:

> Over the noise of Lopez's machine-gun firing I could hear Captain Wilson shouting to withdraw into Rocherath. I wanted to obey, but I was caught in the cross-fire of the heavy machine-gun and the attackers. I gritted my teeth and waited for a lull in the firing. None came. I jumped from the hole and ran blindly toward the rear. Bullets snipped at my heels. The tank saw that we were running again and opened up with renewed vigour, the shells snapping the tops from the trees around us as if they had been matchsticks. I felt like we were helpless little bugs scurrying blindly about now that some man-monster had lifted the log under which we had been hiding. I wondered if it would not be better to be killed and perhaps that would be an end to everything.[23]

Others, like Lieutenant Eric Woods, Executive Officer of the 589th Artillery Battalion's Battery A, and a former Princeton football player, stuck it out until capture was inevitable, and then took to the hills, quite literally, living off looted supplies and carrying on a one-man guerrilla war against German supply columns. Woods was never seen alive again; his body was recovered on 23 January, when the 7th Armoured Division recovered the area, dead at a crossroads in the forest, his Colt .45 automatic empty by his side, and dead Germans all around. He was buried where he lay, beneath a stone which reads:

'Eric Fisher Wood/*fand hier den Heldentod/nach schweren Einzelkämpfen*' ('Eric Fisher Wood died a hero's death after unsparing single-handed combat').

While much of the Ardennes region is covered with forest, there are large reaches of open country, too, and here such resistance as the Americans were able to offer took its toll in a way more reminiscent of the previous world war, and all the more horrific for being totally unexpected:

The open and comparatively smooth slope which separated Mont and Neffe [two villages due east of Bastogne] afforded a [good] field of fire for Griswold's battalion. In addition . . . the slope was criss-crossed with man-made obstacles – a chequer-board of barbed-wire fences erected by the Belgians to make feeder pens for their cattle. The fences were in rows about thirty yards apart. Each fence was five or six strands high. Because of the manner of their construction, it was almost impossible to crawl under-neath the fences; a man approaching Griswold's forces at Mont had to halt at each obstacle and climb through.

Whether the German commanders knew of the existence of these obstacles and decided to risk the attack anyway, or whether the leading [infantrymen] just stumbled on to them by accident in the darkness, will probably never be known. But the attack, once launched, had to be carried on. The German infantrymen ran forward with the same enthusiasm, the same wild yells and eagerness which had characterized all their offensive actions since the breakthrough. When they reached the fences, they simply climbed through. But it broke them.

As fast as they reached the obstacles, Griswold's machine-gunners swept them down. The Germans were in great strength, and the forces behind, pressing upon the forces ahead, made the massacre inevitable. Bodies of the dead piled up around the wire fences, and the attackers who followed, climbing over those bodies, became bodies themselves a few steps beyond. The volume of tracer fire from Griswold's gunners was spectacularly intense; prisoners questioned later said that its visible effect, as much as the holocaust around the fences, was to them the terrifying element of the night attack.[24]

The final assault on the Third Reich in early 1945 was spearheaded by battle-hardened US infantry units which had fought all the way across the Continent from the Normandy bridgehead.

Few of the infantrymen who stormed ashore on D-Day were still serving in the front line six months later due to heavy casualties in the bloody campaign, culminating in the Battle of the Bulge. The Americans used a battle casualty replacement system, where soldiers from a pool were shipped forward to make up numbers as units started to suffer casualties. They had little time to get to know their comrades, only to die beside them in battle. In the space of a few days in the Ardennes Offensive, for example, one US battalion lost 461 soldiers and got 347 replacements.

The twelve-man rifle squad was the smallest unit in the US infantry; it was commanded by a sergeant who more often got his rank as a result of combat promotion because of heavy casualties rather than through training or seniority. Each squad had Browning automatic rifles (BARs) for long-range firepower, two or three Thompson or M3 sub-machine-guns (the famous 'Grease Gun'), and the remainder were armed with M1 Garand rifles. The Garand was the first automatic rifle to be issued *en masse* to a major army, and it provided GIs with long-range sustained firepower. Many sections also boasted a Bazooka anti-tank rocket launcher that could hit targets up to a range of 150 metres. By the later stages of the war, at least one man in each squad had an M1903 bolt-action sniper rifle.

In theory the rifle squad was split into three sections. Able section was a two-man scout team; Baker section was a four-man fire team which included the squad's BAR and sniper rifle; and Charlie section was a five-man manoeuvre and assault team. American fire and manoeuvre squad tactics were complicated, but due to heavy casualties a squad rarely had enough men to operate by the book. Where possible, therefore, squads tried to get hold of extra BARs to try to equalize the odds with German troops using the belt-fed MG42.

American infantry squads faced their toughest test in the hedge-rows of Normandy, where they faced the experienced German *Waffen-SS* and paratrooper units. The Americans ground down the Germans using their superior artillery and tank support; by working closely with support units the GI infantry never had to take on the Germans' best infantry on equal terms.

The weather was an additional enemy to the men besieged in Bastogne, for low cloud and rain prevented both their resupply by air and fighter-bomber aircraft from giving them vital close support. It was 23 December before it cleared and then, in temperatures many degrees below zero, even at midday, fresh supplies of food, weapons and ammunition began to fall out of the sky into a drop zone just a mile square, over 95 per cent of the 1,446 loads delivered being recovered successfully.

Bastogne was relieved on 26 December, but it was a full month more before the Allies had pushed the Germans back to their start-line, and by then, Eisenhower had made his plans for the final assault on Germany. It was to be the bitterest fighting of any seen in the West; the Germans who had fought so hard to hold on to their gains in France, and who were still proving difficult to shift in Italy, found renewed strength when it was their own homeland at stake, even though the weapons facing them were fearsome, and the tactics well honed. They fought for every house, and for every yard of ground:

A Lancer troop moved up on his left, their Besas [machine-guns] spitting tracer. They were going to shoot him in. He felt an intense comradeship with the long Cromwells [cruiser tanks]. They and he were out ahead – the little black arrows on the war maps.

'Co-ax, five hundred, fire!'

His own gunner opened up, and the choking fumes blew back into the turret, sharp and choking, towards the ventilator fan.

'Keep spraying, gunner.'

They moved with infinite slowness. It seemed to be getting lighter, but really it was just the distance closing. The low black line of the trees emerged, the spire of the church, dim against the dawn.

'Get ready, flame-gunner. . . . Fire!'

The flame shot out, fell, broke, rolled along the ground.

'Left, sir! Left!'

Suddenly, from the side of the periscope, Wilson saw something flash against the armour of a Cromwell. The driver jerked the Crocodile towards a small, dark opening in the trees, and for one long moment the flame-gunner pumped the fire.

They worked down the trees which masked the front of the village, pouring the fire into the darkness. Now and again there was the sound like men screaming, which Wilson had once heard in Normandy. When they reached the end of the trees, Wilson halted the troop and they stood off the target while the infantry went in.

He led the Crocodiles into another opening in the trees, and everything went dark again. The front of his tank began to nose up a tall bank; it lifted slowly, reached the top and stood poised for a moment. All at once the bank gave way.

'She's slipping,' shouted the driver. 'I can't hold her.'

One of the tracks started to race. The tank began to turn over, sliding a little, rolling on its side. Wilson thought: We shall be helpless, like an upside-down turtle. Next moment the tank slid off the bank and crashed into a dark space below.

His head must have struck the gun mechanism. When he came to, the tank was on its side. The seventy-five and the Besa were useless. The wireless was dead. All he could see was the red indicator lamp of the flame gun, which still glowed on the turret wall.

'Are you all right in front?'

The flame-gunner answered, sounding dazed.

'Can you see anything?'

'Yes, a house.'

'All right, flame it.'

The flame shot out. Its sudden glow lit up the periscopes. He leant against the seventy-five's ejector shield and tried to manip-ulate them, straining to see the enemy who must be coming in with their Bazookas [*sic*: German troops would have been armed with the rather more powerful *panzerfaust*]. But the periscopes would move only a few degrees.

He directed the flame to the only other target he could find – a group of cottages, a little to the right, about a hundred yards away. The fire crashed in and ran through the buildings from end to end. It's always the same, he thought. Flame everything in sight and you're terrifying. Stop, and you're a sitting target for the Bazookas [*sic*]. But, fired continuously, the flame only lasted for a few minutes . . .

All at once, the gun gave the snort which indicated that its fuel tanks were empty.

'Take your guns and get out,' he said.

They climbed out with their Sten guns and made a small group around the Crocodile. He saw now that, coming over the bank, it had run on to the roof of a small house. The house had collapsed, and the tank had fallen down among the rubble. The battle was going on all around.

'Wait here,' he said. 'I'll try and find the others.'

He took a Sten and went through some bushes to the right . . . [the garden there] seemed quite empty. Then he saw the German officer. He was standing ten yards away, with a pistol in his hand, and he saw Wilson at the same moment.

Wilson pressed the trigger of his Sten. He thought: 'This is the first time I've killed a man this way.' But nothing happened. He tugged at the cocking handle. The Sten was jammed.

Slowly, the German raised his pistol and fired. The bullets smacked dully into the bushes at Wilson's back. It seemed so stupid. He was struggling to get out his own pistol, which was caught in the lining of his pocket.

Then the German turned and ran. As he went through the gate, a Bren gun fired on the far side of the wall: he tilted sideways, and fell in a little heap.

Wilson found the Bren and a corporal in charge of it.

'Have you seen the Crocodiles?'

'Frig, no,' said the corporal. 'But the place is lousy with Jerries.' He was going in to clear the Rectory. His men were all-in. They'd been doubling in through the Spandau fire, and now he was urging them on again.

At last he got them to their feet. They ran across a lawn and the corporal threw a grenade through the door. As they went in, he shot left and right through every door with the big German Schmeisser [*sic*, machine pistol] he was carrying . . .[25]

Captain Andrew Wilson then returned to the battle in a Crocodile commandeered from another troop. In the lull which followed the fighting, sometime after day had broken, he found himself at the edge of the area where the battle had taken place . . .

Ever since his first action in Normandy he'd been drawn by a fierce curiosity to see what happened where they had flamed . . . He walked down the front of the trees, where here and there the brushwood still smouldered among the blackened trunks. The burning away of the undergrowth had completely uncovered some trenches. He looked in the first and for the moment saw only a mass of charred fluff. He wondered what it was, until he remembered that the Germans were always lining their sleeping-places with looted bedding.

Then, as he turned to go, he saw the arm. At first he thought it was the charred and shrivelled crook of a tree root; but when he looked closer, he made out the hump of a body attached to it. A little way away was the shrivelled remains of a boot.

He went on to the next trench, and in that there was no concealing fluff. There were bodies which seemed to have been blown back by the force of the flame and lay in naked, blackened heaps. Others were caught in twisted poses, as if the flame had frozen them. Their clothes had been burnt away. Only their helmets and boots remained . . .[26]

While the British and American armies were fighting their way towards the Rhine, the Russians, in the east, had swarmed across the Ukraine and through Poland, and were now crossing into the Reich itself in a mad rush for Berlin, creating a huge salient. Further south meanwhile, Hungary west of Budapest – garrisoned by 50,000 German and Hungarian troops, a fifth of whom were already wounded – as well as Czechoslovakia and Austria, were still in German hands.

In April, the main strength of the *Wehrmacht* finally failed, and from then on, the Allies were tumbling over each other in their haste to bring the war to its conclusion. By 25 April, the Red Army had formed a ring of steel around central Berlin, composed of an infantry army, two shock armies, a guards army and three guards tank armies – 464,000 men, 12,700 guns and heavy mortars, 21,000 multiple rocket launchers and 1,500 tanks – while the American, British and Soviet air forces bombarded the city night and day from high altitude.

Casualties on both sides for this battle for the Reich's capital were horrific. The sheer weight of fire from the Soviet artillery batteries tore up the streets themselves and crashed through buildings, tearing them down along with the snipers and machine-gun crews infesting them, creating further chaos when the huge Josef Stalin tanks ploughed through the rubble to seek out other nests of fanatical German fighters, many of them sixteen years of age and younger. As the hours passed, the resistance was slowly worn down, until by the evening of 27 April, the German-occupied zone was reduced to a single narrow pocket, 16 kilometres long by 5 kilometres wide, centred on the bend in the River Spree where the Reichstag stood. Yet it took another three full days before the Reichstag itself and the nearby Ministry of the Interior, manned by desperate SS men who had nothing to lose, actually fell to the Red Army. By then Hitler, the architect of it all, and the arch-enemy, was himself dead. The war in Europe was all but over, though it was actually 14 May, ten full days after the preliminary armistice was signed on Lüneburg Heath and eight days after General Jodl signed the Unconditional Surrender at Eisenhower's headquarters at Rheims, that the fighting actually stopped.

127

In Asia, the Americans were still facing strong opposition as they moved closer to the Japanese Home Islands, though with the Philippines now under their overall control and the British forcing their way down through Burma towards the Malay border, it was only a matter of time before Japan was forced to her knees by a simple lack of raw materials and fuel. During the battles for Okinawa, it had become clear that any invasion of the Japanese main islands would be a very costly business indeed, and that the entire population could be expected to fight tooth and nail; even the optimists in the Pentagon were heard to talk of years more of warfare and taking a million further casualties. But it was not to be; the Manhattan Project finally came to fruition with a successful test at Alamogordo, New Mexico, at 05.30 on 16 July 1945, and within three weeks Japan, too, had capitulated after nuclear bombs were dropped first on Hiroshima, and then on Nagasaki. The Second World War was over, leaving a world changed almost out of all recognition behind it.

Chapter Four

Commandos and Special Forces

By the time the Second World War ended, it had become an intensely technological – not to say scientific – affair, but that fact notwithstanding, the war as a whole had probably involved a greater volume of close-quarter fighting than any previous conflict in the history of mankind, if only due to its truly global extent and the time-scale it encompassed. In the beginning, however, and particularly in the dark days of 1940, when everything seemed to be going Germany's way and Britain's back was very firmly against the wall, there was a desperate need both to take the battle to the enemy and to fight back with any means available, and the only solution at hand was to introduce a formalized element of unconventionality into modern warfare.

Armies have employed independent bands of unconventional fighters ever since warfare became organized, no doubt. In comparatively recent times we can nominate Rogers' Rangers during the American War of Independence; Davidoff's Cossacks, who harassed Napoleon's Grand Army all the weary way back from Moscow; the Spanish guerrillas of the Peninsular War; the Boer commandos of the South African War; T.E. Lawrence's flying columns of Arabs and Wingate's Special Night Squads in Palestine, though that list is far from exhaustive.

After the fall of France in June 1940, there was scarce opportunity for the British government to stiffen the nation's will to resist with news of victories, however small. This was of great concern to Prime Minister Winston Churchill, who certainly knew the importance of propaganda in the maintenance of morale, and even while the evacuation from Dunkirk was taking place he was issuing instructions to his Chiefs of Staff to begin gathering and training troops for a new 'unconventional' type of operation, not at all dissimilar in character to the trench raids of the First World War, though potentially bigger and certainly involving greater distances: 'a "butcher-and-bolt" reign of terror', in his own words, to be launched against Hitler's Festung *Europa*.

The new kind of warfare was to be waged by what became known as 'commandos'. (There was significant resistance in traditional circles to this name being given to these new-model soldiers, in that it was the one used by the Boers for the flying columns they had employed in the near-disastrous South African war of just forty years before; it took the intervention of Field Marshal Sir John Dill, the Chief of the Imperial General Staff (CIGS) himself, finally to get the name adopted, and even then the ambiguous and spiritless – though menacing enough – term 'Special Services' – which was also distinctly inappropriate, particularly when reduced to its initials – still managed to creep in from time to time.) In the process, the commandos turned close-quarter combat into a formalized fighting skill and created a whole new sub-class of 'super-tough' fighting men, soon to be acknowledged as the elite forces of the modern army.

In contemporary terms, the decision to set up the new force had immediate precedents: the British Army's ten Independent Companies, raised in the spring of 1940, which were organised with similar tasks in mind, though the level of special training they received was hardly on a par with that of their successors. Recruited mainly from the Territorial (i.e., not Regular Army) Divisions, and each comprising (nominally, at least) 20 officers and 270 men, they were trained and armed principally to carry out sabotage raids on enemy lines of communications. Five of them – and they can hardly have received sufficient training – went to Norway, post-haste, to attempt to stem the Nazi invasion of May 1940, but the rest seem to have been used as 'ordinary' infantrymen during the Battle of France and the subsequent retreat to Dunkirk. What was left of the companies were disbanded in November of that year, though many of the surviving men who served in them went on to see action with other 'special' units. While this is by no means a history of Special Forces – inevitably, the term was soon endowed with initial capital letters, that first step towards military canonization – there is no escaping the role such units played, during the Second World War and since, in any even slightly out-of-the-ordinary form of combat. They are of particular interest to us since many special operations were (and continue to be) necessarily carried out at very close quarters indeed.

In the summer of 1940, a Royal Artillery Colonel named Dudley Clark was serving as Military Assistant to the CIGS, Sir John Dill. After the debacle-turned-miracle of Dunkirk (the most unlikely people turned out to be competent small-boat sailors when the life-blood of the British Army was at stake; Richard Meinertzhagen, for example, whom we last encountered slaughtering Russian agents in Andalusia, made two journeys to the Belgian beaches despite being officially designated Home Guard commander with responsibility for Whitehall and Downing Street!), and in response, no doubt, to Churchill's order to implement 'butcher-and-bolt' operations, Clark reviewed some expedient means of striking back at the enemy, and put before Dill a paper advocating the 'Commando Plan'. It met with Churchill's approval, and Clark was given limited permission to begin recruiting – though was warned off poaching from the newly formed Home Forces Command, virtually all that was left of the British Army in Europe, in fact, and with a very firm priority on men and equipment.

British Commandos were unorthodox in everything they did. In an age of mass conscript armies, they relied on volunteers. While conventional armies relied on firepower, the Commandos used cunning and stealth to defeat the enemy.

The first British Commando unit was officially called the Special Service Brigade. It was made up of battalion-sized units called Commandos; each Commando was made up of troops, which were platoon-sized units of around thirty men. Every Commando was a volunteer who had seen active service in the field army before moving over to the new raiding force. Outwardly, discipline appeared lax compared to regular units, but the training was intense and physically demanding. Failure to meet standards in training would result in budding Commandos being sent back to their parent units in disgrace.

Unlike in conventional units, Commandos were trained in map reading, night navigation and leadership, so they could complete their unit's mission if officers or NCOs were killed in action. Every member of the Commando units had to be fully briefed on every aspect of a raiding plan so they could use their initiative to overcome enemy resistance if things started to go wrong – they usually did!

Mobile firepower was the key to Commando success. Sea-landed raiding troops were liberally supplied with .303 Bren light machine-guns, American Thompson and British Sten sub-machine-guns and American M1 Garand automatic rifles. The Commandos carried rucksacks full of ammunition rather than food or sleeping bags. Three-inch mortars were carried for heavy firepower, and every member of a raiding party would have to carry the mortar's bombs, dropping them off at the base plate position before moving off to launch the attack. When the Commandos opened fire, they did so from concealed positions at night so the Germans had little idea what was happening until the raiders were among them.

Commando raiders would then set about their work of destroying the target with demolition charges. Once they were set, the Commandos would withdraw to a safe distance before blowing the explosives by remote control or with timing devices. The with-drawal was often the most difficult part of a raid. Commandos were given prearranged rendezvous points to meet at before they would move as a group to reembark on their landing craft. If they missed the boat Commandos were on their own. They could either surrender or escape to friendly territory. Few Commandos chose the first option.

Author William Seymour, who himself served with 52 (Middle East) Commando in 1941–42, during the course of a fifteen-year career with the Scots Guards, described the requirements for a Commando soldier in his comprehensive history, *British Special Forces*:

A good Commando soldier had to be a protean figure. Every worthwhile soldier needs to have courage, physical fitness and self-discipline, but those who served in the Commandos needed these virtues more abundantly, for they would be called upon to perform feats well beyond the normal run of duty, and often to work longer hours and enjoy less rest than their counterparts in a regiment. A high standard of marksmanship had to be attained, as did the ability to cross any type of country quickly, to survive a rough passage at sea in a small boat and arrive [at] the other side ready to scale a cliff, to use explosives, to think quickly and if needs be to act independently, and to be prepared to kill ruth-lessly. A man would learn some of these things in his training, but vitality and a zest for adventure must be the well spring of every one of his actions.[1]

In fact, there was every good reason to suggest that Seymour actually had his priorities reversed, and was subscribing in that last sentence to an elitist myth, when necessity called for an altogether more down-to-earth recruiting programme. A man might well have vitality and a zest for adventure of the necessary sort – perhaps one man in a thousand might, if the truth be known, male vanity aside – but without proper training, he would never amount to the type of skilled fighting man the new task required. One in a thousand simply would not be enough, in any event, because the same sort of adventurous spirit was still required in the junior leaders of conventional forces, too. What was needed was a training course designed to turn little more than ordinary, averagely fit, averagely intelligent young men into professional killers, and the first priority was to select the men who would design and run it.

It has been suggested that one of the strengths of the British Empire was the presence, somewhere in the ranks of its administrators, of one or more men capable of carrying out literally any task under the sun, and this one, though odd in the extreme, proved to be no exception.

William Fairbairn, universally known as Dan, and Eric Sykes who, in a world familiar with the works of Charles Dickens, was invariably called Bill, had served together for over two decades in one of the roughest, toughest milieus known to man: the city of Shanghai. Fairbairn latterly was Assistant Commissioner, and head of the city's riot police, and Sykes was head of the sniper unit and chief firearms instructor. Fairbairn was the first foreigner living outside Japan to be awarded a Black Belt in ju-jitsu, and was an acknowledged expert in unarmed combat of all sorts – 'practically every known method of attack and defence', as the preamble to his *All-in Fighting* of 1942 has it. The two were recalled from China in 1940, commissioned as captains in the British Army, and subsequently these nondescript, middle-aged men, both of them sporting neat moustaches, portly Sykes and bespectacled Fairbairn (who had, according to a colleague, been personally involved in over 200 close-combat incidents), turned lethal hand-to-hand combat into a formalized fighting skill, accessible to all, based on moves developed 'in dealing with the thugs, ruffians, bandits and bullies of one of the toughest waterfront areas in the world'.

Both men were true believers in the power of the knife: 'At close quarters, there is no more deadly weapon than the knife,' Fairbairn was in the habit of saying as preface to his introductory lecture on the subject, 'and it never runs out of ammunition.' In order to introduce the knife as a fighting weapon, however, it was first necessary to come up with – and put into production – a suitable design, for none existed. Fairbairn and Sykes, while heavily involved in developing the overall curriculum for the new close-quarter battle school, produced a set of very basic criteria for a slashing and stabbing weapon: it should have a heavy, roughened grip, a stout cross-guard, and a double-edged symmetrical stiletto blade, just under eighteen centimetres in length. They approached the Wilkinson company which produced the British Army's swords, and they agreed to put the Fairbairn-Sykes Fighting Knife, as it became known, into production at a cost, to the British Army, of 13/6 (thirteen shillings and sixpence or 67.5p each), including sheath. Before the war ended, Wilkinson Sword had manufactured over 250,000 of them, though it's true that most served no more violent purpose than to open recalcitrant tins of bully beef.

By the time the first models of the new weapon became available, Fairbairn and Sykes were teaching their close-quarter combat course at the Special Operations Executive's 'Experimental Station 6' – Aston House, near Stevenage in Hertfordshire – where Sykes was 'the master of soundless strangulation', and Fairbairn demonstrated a variety of disabling techniques which had very little to do with ju-jitsu. Now they expanded their instruction to embrace knife-fighting, too. Other 'killer schools' served other branches. The SOE (Special Operations Executive) had many, including Hatherup Castle in Gloucestershire; the Commandos proper found a home on the Duke of Argyll's estate at Inverary and later organized the most famous of the special forces schools at Achnacarry, as well as the Mountain and Snow Warfare Camp at Braemar, all in Scotland. March-Phillips's Small-Scale Raiding Force, later redesignated 62 Commando, had Anderson Manor, in Dorset, and later Luton House, near Torquay, as well as Wraxall Manor, near Crewkerne. The Mediterranean-based Special Boat Section had one establishment in Malta and another in Jerusalem, and later also acquired one at Ardrossan, in Scotland, when it turned its attention on north-west Europe. The Special Air Service, which also came into being in North Africa, had a number of training bases there, and one at Athlit, in Palestine, and later also resorted to the wilds of Scotland. In all of them, lethal hand-to-hand fighting formed an important part of the curriculum.

'The knife in close quarters fighting is the most deadly weapon to have to contend with,' said Fairbairn in a summary of his and Sykes's training doctrine, published as *All-in Fighting*, in 1942 (each page of instruction of which is accompanied by a facing page of illustrations in the form of line-drawings, so as to leave nothing to the imagination). The authors' continue:

It is admitted by recognized authorities that for an entirely unarmed man there is no certain defence against a knife. With this we are in entire agreement. We are also aware of the psychological effect that the sudden flashing of a knife will have on the majority of persons.

It has been proved that the British bayonet is still feared, [but] it is not very difficult to visualize the many occasions, such as on a night raid, house-to-house fighting, or even a boarding party, when a knife or short broadsword would have been a far more effective weapon.

There are many positions in which the knife can be carried . . . This is a matter which must be decided by each individual for himself . . . We personally favour a concealed position, using the left hand, well knowing that in close-quarter fighting the element of surprise is the main factor of success . . . Speed on the draw can only be acquired by daily practice.

It is essential that your knife should have a sharp stabbing point, with good cutting edges, because an artery torn through (as against a clean cut) tends to contract and stop the bleeding. This frequently happens in an explosion. A person may have an arm or a leg blown off and still live, yet if a main artery had been cut they would quickly have lost consciousness and almost immediately have died.

Certain arteries are more vulnerable to attack than others, on account of their being nearer the surface of the skin, or not being protected by clothing or equipment. Don't bother about their names as long as you remember where they are situated . . . They vary in size from the thickness of one's thumb to [that of] an ordinary pencil. Naturally, the speed at which loss of consciousness or death takes place will depend upon the size of the artery cut.

The heart or stomach, when not protected by equipment, should be attacked. The psychological effect of even a slight wound in the stomach is a point worthy of note.[2]

Initially, Fairbairn and Sykes made some basic tactical mistakes in specifying how the knife was to be used in surprise attacks on German soldiers – the junction of the heavy leather braces which were a feature of *Wehrmacht* battledress, for example, protected and shielded a vulnerable point in the spinal column, while other items of kit obstructed the kidneys – and as a result those particular targets were eliminated. The prime objective, however, when trying to silence a sentry from behind, was always the subclavian artery, which is located within the soft tissue at the junction of neck and shoulder, behind the collar-bone; a single powerful, downward thrust through the target zone is usually enough to sever it, and the result is unconsciousness in two seconds, and death very soon afterwards. Fairbairn taught his pupils to approach from behind, hook a hand round the victim's chin and wrench his head around, thus exposing the target area to the knife. 'This is not an easy artery to cut,' states *All-in Fighting*, 'but once cut, your opponent will drop, and no tourniquet or any help of man can save him.' While the techniques of knife fighting – and especially, perhaps, the lethal attack on an unwary opponent – were straightforward in theory, it took a very determined man to carry the manoeuvres through to their fatal conclusion, as some of Fairbairn's 'students' were to discover. And many of those who were called upon to put their training into practice were horrified at the actuality of it: 'Killing a sentry silently usually required two men,' a Second World War vintage Commando recalled, not without very considerable distaste and unease, even a long time after the event. 'One holding, the other knifing. It's a hateful job. I don't like to think about that. It's a sad story with me, the filthiest fighting I know. For months after the war, when what I had done came back to me, I had to take pills to help me overcome the remorse.' And that, of course, was the inevitable side-effect of making professional killers out of ordinary young men, though in fact, very few were actually called upon to use the knife in the way described. Fairbairn, who seems to have had either few human feelings or an iron will, offers but a single crumb of sympathetic advice to the faint-heart in the entire book; at the very end of the section on hand-to-hand fighting he says: 'If you are inclined to think these methods are "not cricket", remember that Hitler does not play this game.' None the less, it is a very difficult thing, even in a time of all-out war, to steel oneself to committing cold-blooded homicide upon an unwitting subject.

Some men undoubtedly either had the clearness of vision to

understand the necessity for such acts, or perhaps lacked the imagination or moral sense necessary to see them as inhuman, though as we have had cause to note before, within the context of all-out war such notions as mercy and humanity towards one's enemy are potentially very dangerous indeed on a personal level, where the choice is often a very stark kill or be killed. Thankfully, most close-combat situations are anything but cold blooded, and a completely different set of rules and criteria applies.

Fairbairn and Sykes did not limit themselves to the design of just one type of fighting knife, though the commando dagger, as the No 1 Fighting Knife came to be widely known, was certainly their most famous creation. Their No 2 knife, designed the following year, was an altogether different type of weapon, perhaps inspired by the 'short broadsword' they referred to, though it also has a certain similarity with the Gurkha kukri. The 'smatchet', as the knife was christened, was much bigger and heavier than the dagger – a chopping weapon, predominantly, though it also had a sharp stabbing point. Its leaf-shaped blade was some thirty centimetres long, and ten broad at its widest point, and the grip, separated from the blade by a wide brass crosspiece, incorporated an exaggerated brass pommel which was itself designed to be used as a secondary weapon. The smatchet weighed slightly in excess of one kilogramme. It was certainly a menacing-looking tool, and instinct leads one to suspect that it would have proved very popular to an earlier generation of trench raiders, but during the Second World War it failed to live up to its creators' hopes. They said of it: 'The psychological reaction of any man, when he first takes the smatchet in his hand, is full justification for its recommendation as a fighting weapon. He will immediately register all the essential qualities of a good soldier – confidence, determination and aggressiveness.'[3]

Its balance, weight and killing power, with the point, edge or pommel, combined with the extremely simple training necessary to become efficient in its use, make it the ideal personal weapon for all those not armed with a rifle and bayonet.

Tips for its employment included, 'Drive well in to the stomach', 'Sabre cut to the base of the neck, the wrist or the inside of the elbow' and how to land an upwards or downwards blow with the pommel to the jaw or chin.

All-in Fighting did not restrict itself to unarmed combat and use of the fighting knives; indeed, the short section on attacking one's

137

unsuspecting opponent with a short stick or cane is perhaps the most frightening in the entire small volume, for each movement has the potential to be lethal, and with a weapon which many would consider to be no weapon at all. In his introduction to that section, Fairbairn comes as close as he ever does to displaying the wider worth of his teachings, however: to instil confidence and boost morale:

> A man without a weapon to defend himself, especially after long exposure, is very liable to give up in despair. It is remarkable what a difference it would make in his morale if he had a small stick or cane [18 – 24 inches long and one inch in thickness is Fairbairn's recommendation] in his hand. Now, add to this the knowledge that he could, with ease, kill any opponent with a stick and you will then see how easy it is to cultivate the offensive spirit which is so essential in present-day warfare.[4]

I have no intention whatsoever of reproducing any of Fairbairn's attacking moves with a small stick, or indeed of going into detail of such ploys as the celebrated 'matchbox attack', of which Fairbairn says: 'The odds of knocking your opponent unconscious by this method are at least two to one. The fact that this can be accomplished with a matchbox is not well known, and for this reason is not likely to raise your opponent's suspicions ... Naturally, all movements, from the initial start of the blow, must be carried out with the quickest possible speed'; the 'weapons' in question are far too innocuous and easily available, while the moves are both straightforward and easy to learn, and undoubtedly lethal.

The preface to *All-in Fighting*, written by Lieutenant-Colonel J.P. O'Brien Twohig, sums up not just the overall purpose of such training, but also addresses something more resembling the reasons for setting up the special forces in the first place, given that their contribution to the war as a whole was actually seldom more than of nuisance value: 'The principal value ... lies not so much in the actually physical holds or breaks, but in the psychological reaction which engenders and fosters the necessary attitude of mind which refuses to admit defeat and is determined to achieve victory.'

The introduction, on the other hand, justifies the need for the book (and also for training men according to its principles) in very clear terms:

Some readers may be appalled at the suggestion that it should be necessary for human beings of the twentieth century to revert to the grim brutality of the Stone Age in order to be able to live. But it must be realized that, when dealing with an utterly ruthless enemy who had clearly expressed his intention of wiping this nation out of existence, there is no room for any scruple or compunction about the methods to be employed in preventing him. The reader is requested to imagine that he himself has been wantonly attacked by a thug ... Let him be quite honest and realize what his feelings would be. His one, violent desire would be to do the thug the most damage – regardless of rules. In circumstances such as this he is forced back to quite primitive reactions ...

There are very few men who would not fire back if they were attacked by a man with a gun, and they would have no regrets if their bullet found its mark. But suggest that they retaliate with a knife, or with any of the follow-up methods explained in this manual, and the majority would shrink from using such uncivilized or un-British methods. A gun is an impersonal weapon and kills cleanly and decently [sic] at a distance. Killing with the bare hands at close quarters savours too much of pure savagery for most people. They would hesitate to attempt it. But never was the catchword [sic] 'He who hesitates is lost' more applicable. When it is a matter of life or death, not only of the individual but indeed of the nation, squeamish scruples are out of place. The sooner we realize that fact, the sooner we shall be fitted to face the grim and ruthless realities of total warfare.[5]

Sinew-stiffening stuff indeed.

Fairbairn and Sykes also turned their attention to the selection and use of firearms, especially pistols, and once again offered sound advice to the beginner by organizing the lectures given to 'Commando School' classes in Britain and the United States in book form, and publishing them as *Shooting to Live with the One-hand Gun*. The two clearly knew what they were talking about – the records of the Shanghai Municipal Police (SMP), which they used as an informal database to support their theories, over twelve and a half years between 1928 and 1940 referred to 666 armed encounters with criminals. In those in which pistols were used by the police, 42 officers and 260 criminals were killed, 100 officers and 193 criminals were wounded.

'Excluding duelling (since it is forbidden in most countries and appears to be declining in popularity in those countries in which it is permitted),' they begin:

> there seem to remain two primary and distinct uses for the pistol. The first of these . . . is . . . target shooting. Its secondary use is as a weapon of combat.
>
> In the great majority of shooting affrays the distance at which firing takes place is not more than four yards. Very frequently it is considerably less. Often the only warning of what is about to take place is a suspicious movement of an opponent's [sic] hand. Again, your opponent is quite likely to be on the move. It may happen, too, that you have been running in order to overtake him. If you have had reason to believe that shooting is likely, you will be keyed-up to the highest pitch and will be grasping your pistol with almost convulsive force. If you have to fire, your instinct will be to do so as quickly as possible, and you will probably do it with a bent arm, possibly even from the level of the hip. The whole affair may take place in bad light or none at all . . . It may be that a bullet whizzes past you and that you will experience the momentary stupefaction which is due to the shock of the explosion at very short range of the shot just fired by your opponent – a very different feeling, we can assure you, from that experienced when you are standing behind or alongside a pistol that is being fired. Finally, you may find that you have to shoot from some awkward position, not necessarily even while you are on your feet . . . and if you take much longer than a third of a second to fire your first shot, you will not be the one to tell about it.[6]

On the subject of choosing a type of pistol, Fairbairn and Sykes were at pains to remain objective, and in the process, demolished the myth (promulgated in no small measure by the Colt Firearms Co) that a heavy, relatively slow-moving bullet was necessary to be sure of stopping a man:

> The type of pistol to be chosen depends on the use to which it is to be put. A pistol that meets the needs of the detective or plain-clothes man, for instance, is not necessarily suitable for individual self-defence or for the uniformed serviceman . . .
>
> For a weapon to be carried openly by . . . officers and men of the fighting services, we unhesitatingly avow our preference for

the automatic pistol ... We are familiar with the criticisms so often made of the automatic pistol. It is said that it is unreliable, will often jamb [sic] without provocation and will certainly do so if mud, sand or water gets into the mechanism, and above all, it is not safe.

There have been and possibly still are automatics like that, but one is not obliged to use them.

Apart from the question of reliability, we have found that in comparison with the revolver, the automatic offers the following advantages: it is easier and quicker to recharge; it can be fired at far greater speed and it is easier to shoot with [curiously, they don't mention a factor which modern authorities would agree to be at least as important as any of the foregoing, in the circumstances: the automatic's greater ammunition capacity].

We have an inveterate dislike of the profusion of safety devices with which all automatic pistols are regularly equipped. We believe them to be the cause of more accidents than anything else. There are too many instances of men being shot by accident either because the safety catch was in the firing position when it ought not to have been or because it was in the safe position when that was the last thing to be desired. It is better, we think, to make the pistol permanently 'unsafe' and then to devise such methods of handling it that there will be no accidents ... Our unorthodox methods have been subjected to the acid test of many years of particularly exacting conditions and have not been found wanting.

We have to admit that in the beginning we paid little attention to the magazines or their condition. We soon noticed, however, that some of the magazines in our charge were getting rusty and that others, if not rusty, were clogged up with tobacco dust, fluff and bits of matches, the sort of thing which is found in most men's pockets ...

Enough has been said to show that the condition of the magazine is of the utmost importance to the reliable functioning of the pistol and that at least ordinary care, therefore, should be exercised in its regard. Those individuals who use their magazines as screwdrivers, or to open beer bottles, have no one to blame but themselves when their pistols refuse to function.[7]

They now turn their attention to the choice of ammunition calibre and type:

We are not greatly in favour of small weapons. No small weapon can possess the strength and reliability of a large one.

We were brought up to believe that a heavy bullet of soft lead, travelling in the leisurely manner of bygone days, could not be improved upon if it was desired to dispose of one's foes in a decisive and clean-cut manner. We believed that such a bullet would mushroom, and that even if it did not do so, the impact of such a formidable mass of lead would infallibly do all that was required, including knocking the enemy clean off his feet.

We also believed that bullets of approximately equal weight, jacketed with cupro-nickel and travelling at perhaps a greater velocity, provided penetration as opposed to shock and were therefore unsuited to their purpose; and we had no faith whatever in light bullets driven at a much higher velocity, unless they could be so made as to secure [sic] effective expansion shortly after impact. Expanding bullets [that is, hollow-points or those with incisions in the projectile, usually known as 'dum-dums'], however, are barred by the rules of the game as we have had to play it, so for practical purposes we must confine ourselves to solid bullets.

We are not so sure now of these beliefs. Perhaps the reasons for our doubts will be more easily apparent if we recount some actual experiences from the long list of our [SMP] records:

A Sikh constable fired six shots from his .455 Webley at an armed criminal of whom he was in pursuit, registering five hits. The criminal continued to run, and so did the Sikh, the latter clinching the matter finally by battering in the back of the criminal's head with the butt of his revolver. Subsequent investigations showed that one bullet only, and that barely deformed, remained in the body, the other four having passed clean through.

A European patrol sergeant, hearing shouts of 'Ch'iang-Tao' (robber), rushed to a rice shop which seemed to be the centre of the tumult and there saw an armed Chinese robbing the till. The Chinese immediately opened fire on the sergeant with an automatic pistol, firing several shots until his pistol jammed. Fortunately, none of the shots took effect and meanwhile the sergeant returned the fire swiftly and effectively with a .45 Colt automatic, commencing at about ten feet and firing his sixth and last shot at three feet as he rapidly closed in on his opponent. Later it was found that of those six shots, four had struck fleshy parts of the body, passing clean through, while one bullet remained in the shoulder and another had lodged near the heart. Yet, in spite of all this, the robber was still on his feet and was knocked unconscious by the butt of the sergeant's pistol as he was attempting to escape by climbing over the counter. [Fairbairn and Sykes note that both these accounts refer to jacketed ammunition.]

Turning now to the high-velocity small-calibre weapons, we have seen terrible damage done by a Mauser automatic, calibre 7.63mm, of military pattern. We have in mind the case of a man who was hit in the arm by a solid full-jacketed bullet from a weapon of this type. Though he was in hospital within half an hour of being shot, nothing could be done to avoid amputation, so badly were the bone and tissue lacerated. Perhaps 'pulped' would convey our meaning more exactly . . . Nothing is so feared, rightly or wrongly [by the SMP], as the Mauser military automatic. The mention of the name is sufficient, if there is trouble afoot, to send men in instant search of bullet-proof equipment.

We have tried to solve by experiment this question of the knock-down blow, but there is no satisfactory way of doing it. The nearest we have come to it has been to allow ourselves to be shot at while holding a bullet-proof shield. The chief value of that experiment was a conclusive demonstration of the efficacy of the shield . . . The shock of impact increased in proportion to the velocity of the bullets but in all cases was negligible, the supporting arm only recoiling minutely.

We do not know that a big soft lead bullet will not have the knock-down effect generally claimed. All we can say is that we have never seen it. We do not know for certain, either, that a full-jacketed high-velocity small-calibre bullet will always have the effect described in the particular instance which we have given.

We incline to the belief that the human factor must influence to some extent the behaviour of bullets. A pugilist at the top of his form can stand vastly more punishment that a man who is 'soft' and untrained. Capacity to resist shock and pain appears to be also a function of the nervous system and marked differences occur in this respect as between individuals of different races. Perhaps that partially explains why some men are not knocked out by bullets when they ought to be. Again, if a bullet caught a man off balance, might not that aid in producing the appearance of a knock-down blow?

We have made no mention yet of an aspect of this matter

which we have observed time after time in the course of years. A hit in the abdominal region almost invariably causes a man to drop anything he may have in his hands and to clutch his stomach convulsively. We may add that such a hit almost always has fatal results.[8]

For those with the need to use a firearm at close quarters, but without the wish to become proficient with the pistol, Fairbairn and Sykes, having commenced the chapter on choosing a gun with the stern caveat that, 'Without an adequate knowledge of its use, there can be few things so purposeless and dangerous as a pistol' recommend him: 'to acquire a "sawn-off" shot gun with external hammers of the rebounding type and barrels of about 18 inches in length. The ease with which it can be manipulated, the accuracy with which it can be aimed, either from the shoulder or from the hip, and the spread of the shot charge combine to make it a much safer and more efficient weapon than any kind of one-handed gun in the use of which he is not proficient.' We may note that they do not advocate cutting down the gun's butt into a pistol-type grip.

While Fairbairn and Sykes quite literally 'wrote the book' on hand-to-hand combat, and their work went largely unchallenged (at least in writing), there were others with strong opinions about how firearms training should be organized, not least among them, according to the first generation of SAS soldiers, an American with a commission in the Royal West Kent Regiment, Major L.H. Grant Taylor, also 'portly, nondescript and looking about fifty', whose trademark was a pair of ivory-handled long-barrelled .38 revolvers. 'He restored my belief in cowboys,' said one of his trainees, David Williams:

From ten yards, Grant Taylor would put six bullets through the middle of . . . a playing card. He used to tell us: 'To shoot like me, you have to be born gifted – but you can get somewhere close with practice. His method was: right foot forward, go into the crouch, point the gun as you would your finger, drop it about four inches and pull the trigger.

He taught us to fire left-handed as well as right, and with both hands together. He taught us to fire low, more points being awarded for shots in the body than in the head. A few of us, after being instructed by him, could hit a tin at fifty yards, firing a pistol from the crouch. But we never, ever came near to emulating his other speciality of flicking a coin in the air then drawing from his shoulder holster and hitting the coin before it fell.[9]

The one major difference between Grant Taylor's training methods and those of Fairbairn and Sykes lay in the former's insistence on extensive live firing – where Fairbairn and Sykes usually allocated twelve rounds a day to each pupil, Taylor would have them get through five or ten times that:

'On the course,' said John Lapraik, another pioneering special forces soldier, and one who went on to be Honorary Colonel of 21 Regiment, SAS, 'each of us fired 500 rounds and more. Taylor's method was based on ammunition. Fire and correct, fire and correct. You cannot train people unless they fire ammunition. He required at his passing-out parade that we hit a playing card at 25 yards. Nicking it was enough. And we were expected to do it five times out of six with the right hand, three times out of six with the left.'[10]

As well as marksmanship, firearms training also dealt with the employment of guns, and instructors, such as Sergeant Douglas Howard of the Royal Marines, couched their invaluable advice to trainees in very simple terms, leaving little or nothing to chance: 'When you burst into a room full of enemy soldiers, you must remember the drill evolved for such occasions. Shoot the first man who moves, hostile or not. His brain has recovered from the shock of seeing you there with a gun. Therefore, he is dangerous. Next, shoot the man nearest to you. He is in the best position to cause trouble. Deal with the rest as you think fit.' They also worked hard to simulate realistic close-quarter battle conditions:

'The [Killer] School [for the embryonic Special Boat Squadron, which later included two noted authors, John Lodwick and Eric Newby], was in the old police station in Jerusalem,' recalled Sammy Trafford, another early recruit to the embryonic 'special forces', whom we will soon meet in action on the island of Santorini:

The usual intake was thirty men, taken equally from the SAS, the Military Police and the Palestine Police. We spent every day in physical training and shooting with sub-machine-guns and pistols of our own choice. My favourites were the .38 S&W [Smith & Wesson] and the Thompson ... We didn't use drums; they tended to jam. We used magazines and kept the gun set at auto, although to save ammunition we fired single shots, which required fast reactions and a technique of tapping the trigger instead of pulling it. Anyone who fired two shots instead of one was too slow for the SAS.

We were taught to fire from the battle crouch, rather like a boxer's stance when weaving forward. It was impressed on us that it wasn't necessary to delay fire until our weapon was in the correct, upright position; the barrel remains round and would hit anything it was pointed at no matter which way the gun might have turned in our hands. We worked hard on firing our revolvers with either hand because of the way that doors open. A door hinged on the right and opening inwards called for an entrance with a gun in your left hand. Hinged left and opening inwards meant firing with the right hand. As we left the room, we had to crouch down and reload. I don't believe the scenes on TV of men entering rooms with their gun clasped in both hands while they stand upright like target shooters. Our target was always the big one, captured in this slogan: 'Aim for his guts and he's surely dead.' Not many men recover from a stomach wound.

The tommy-gun magazine holds twenty bullets but we were allowed only ten shots for the passing-out test at the Killer School. Ten bullets, ten targets – and each target popping out suddenly and unexpectedly from cupboards and corners or on stair landings. And, of course, the gun was on automatic. But that wasn't all . . . each examinee was sent round an obstacle course three or four times [first] so that he arrived for the test out of breath . . . I finished joint top with David Williams. We each had 286 marks out of a maximum 300; I'd been better with the tommy gun, he had beaten me with the pistol.[11]

The commandos didn't have long to wait before putting their newly acquired killing skills into practice, thanks in no small part to the basically aggressive nature of the men who were to send or lead them into combat, men like David Stirling, Paddy Mayne, George Jellicoe, David Sutherland, Geoffrey Appleyard, Grahame Hayes, Gus March-Phillips and the Dane, Anders Lassen, and the unceasing demands from on high to be seen to be striking back at the Germans in some way or other, no matter how apparently insignificant.

The first commando raids – they were really ad hoc affairs, and were neither properly prepared nor well executed, though they served Churchill's propaganda purposes well enough – were centred on an airfield at Le Touquet, and on the island of Guernsey, and were carried out in the weeks following Operation Dynamo (the evacuation from Dunkirk) by men of No 11 Independent Company (confusingly, a halfway-house between the ten original Independent Companies and the Commandos proper), but the first major commando operation did not take place until March 1941, when 500 men from Nos 3 and 4 Commandos, 52 Royal Engineers and a like number of men from the Free Norwegian Forces raided the Lofoten Islands, destroying over 3.6 million litres (800,000 gallons) of oil and petrol, sinking 20,321 tonnes of shipping, capturing 216 Germans and 60 Norwegian collaborators and bringing back 315 Norwegian volunteers, at the cost of just one raider wounded – a British officer who shot himself in the thigh. Operation Archery, a raid in similar strength on Vaagso Island at the end of the year, proved equally successful, though this time the fighting, most of it in the streets of South Vaagso, was much heavier. In all 120 Germans were killed while the commandos suffered 17 dead and 53 wounded. Where the Lofoten raid had encountered little opposition, at Vaagso it was patchy – stiff in some places, almost non-existent in others. The Operational Commander, Lieutenant-Colonel John Durnford-Slater, was full of praise both for the way his own men reacted, and for the support accorded them by the ships of the Royal Navy:

I could see everything which took place at Maaloy [Island, where there was a potentially lethal battery of coastal artillery]. Nos 5 and 6 Troops, only fifty yards from the beach when the naval barrage [from HMSs *Kenya*, *Onslow* and *Oribi*] lifted, were up the slopes of the island like a flash. Jack Churchill, who played them in with his bagpipes, was leading them with considerable dash. On landing, Peter Young saw a German running back to man his gun position. 'I was able to shoot him,' Peter told me later. Ten minutes after this, Young reached the Battery Office on Maaloy. One of the company clerks made the literally fatal mistake of trying to wrest Peter's rifle from him.[12]

Peter Young, then a captain but later a brigadier general, won the Military Cross for his part in the fighting at Vaagso, and later distinguished himself both at Dieppe and during Operation Over-lord. When his men had overrun the Maaloy Battery, he hurried them off to South Vaagso, to assist in the action there. Faced with the necessity to clear a large building with a limited number of men, he resorted to a simpler solution, fire:

> The Red Warehouse lay about sixty yards ahead. As we rushed forward through the snow, a German suddenly appeared in the doorway and shied grenades at us. The first two missed, and the third failed to go off. Lance-Sergeants Herbert and Connolly, two of my bravest NCOs, came up and threw at least a dozen grenades into the building. I went in, confident that the occupants must be dead; two shots rang out from an inner room. I left before they could reload. We could find no other way in. What was the answer? I looked at the massive red walls. By God, they were wood! Fire! We must set the place on fire. While I was organizing this, O'Flaherty and Sherington, exasperated by the long delay, staged another assault on the front door. I ran after them and had reached the bottom of the stairs when there were rifle shots; they were both hit. I fired towards the flashes and withdrew, unscathed. O'Flaherty staggered out behind me, his face covered in blood; Sherington had been hit in the leg. I sent them to the rear. A moment later, the warehouse was ablaze. The Germans still refused to surrender and so we pushed on, leaving Lance-Corporal Fison to besiege the burning remains. When it got stuffy, the Germans came out and Fison shot them. We had given them every chance.[13]

The next major operation, which took place in the early hours of 27 March 1942, directed against the port of St Nazaire, was on a slightly smaller scale, in terms of the number of men committed, but it was a true combined operation involving naval forces not just to transport the commandos and give covering fire support but also to participate directly by ramming the gates of the largest drydock in the world, the Forme Ecluse (also known as the Normandie Dock, after the great transatlantic liner it had been built to service), with an explosive-packed destroyer, the First World War-vintage HMS *Campbeltown* (formerly USS *Buchanan*) to prevent its possible use by the battleship *Tirpitz*. It is an indication of the level of personal commitment to the success of the raid that Victoria Crosses were awarded to no fewer than five of the men who participated. The cost was high, too – two-thirds of the raiding party failed to return, 153 of them being taken prisoner. Among the enduring images is that of the ageing HMS *Campbeltown*, launched in 1919 at Bath, Maine, and transferred to the Royal Navy as part of Lend-Lease in 1940, rushing towards her own destruction:

Campbeltown was going fast now, making a good twenty knots, her bow wave splaying wide. Every German gun that could bear was converged upon her; not now those lower down the river, but those at point-blank range in the dockyard itself. From the Old Mole, from either side of the Normandie Dock, from the top of the submarine pens, from the roofs of buildings – and from the east bank of the river. Repeatedly hit, she was now suffering very heavy casualties among her sailors and soldiers alike, her decks spattered with fallen bodies. But miraculously, *Campbeltown* escaped damage to any vital part as she raced forward. Not 'til he was within 200 yards did Beattie see the great steel gate of the dock, discernible as an indistinct black line beyond the 'spill' of the searchlights, dead ahead. At the last moment an incendiary bomb, conceivably dropped from one of our own aircraft, landed on the forecastle and burst into flames. *Campbeltown* held steadily to her course. In the wheelhouse there was complete silence. Beattie and Montgomery propped themselves against the front of the wheelhouse, ready for the shock of impact. Throughout the ship, every man braced himself.

Denser and denser grew the black line ahead. A momentary check told them that they had ripped through the anti-torpedo nets. All hands could feel the wire dragging along the ship's bottom. The caisson was fifty yards away, and only now could Beattie clearly see it. With extraordinary presence of mind, in order to hit the caisson in the centre, and swing the ship's stern to starboard, so that the Old Entrance should not be blocked, he crisply ordered at this last moment:

'Port 20!'

Instantly Tibbits obeyed, and at 01.34, with all her Oerlikons blazing at the enemy guns only a few yards away, with her fo'c'sle in flames, she crashed into the caisson.

She struck with such accuracy and force that her bow, up to the level of the caisson, crumpled back for a distance of thirty-six feet, leaving her fo'c'sle deck, which was higher than the caisson, actually projecting a foot beyond the inner face.

Beattie turned to Montgomery with a smile, and said, 'Well, there we are.' Looking at his watch, he added: 'Four minutes late.'[14]

The precision with which her captain handled the 1,117-tonne, 90-metre-long destroyer ensured that the dock gates were destroyed, while the similar precision with which the shore party placed explosives within the dock's operating machinery ensured that it, too, was past repair. In fact, it was 1950 before the huge drydock was back in operation again.

Such examples of entire ships going into close-quarter battle are rare in the twentieth century (though there were certainly other incidents of ramming; it was particularly effective against submarines), but in an earlier age they often had, for boarding was long the accepted way of subduing an enemy. A ship of the Royal Navy only once sent out an armed boarding party under combat conditions during the Second World War: on St Valentine's Day, 1940, sailors from the destroyer HMS *Cossack*, under the illustrious Captain Philip Vian, boarded the German tanker-turned-prison ship *Altmark* in Jossingfjord, and after a brief struggle in which just one German sailor was killed, released 299 prisoners taken by the commerce raider *Graf Spee* in the South Atlantic prior to her fight with the cruisers *Achilles*, *Ajax* and *Exeter*. Though other enemy ships were certainly boarded by commerce raiders' crews before being sunk, they had all previously surrendered. And then, on the greatest scale yet, came the raid on the port of Dieppe, designed to discover if it was possible to take and hold a French port, as a precursor to invasion. 'A reconnaissance in force', Winston Churchill called it, though there were the usual subsidiary objectives: the destruction of installations and defences, the sinking of enemy shipping and the taking of prisoners. And seen in the light of an experiment, it was a success, though a costly one, if only because it allowed Allied invasion planners to discount the possibility of taking a major port – Le Havre, say, or Cherbourg, or even Calais – and concentrate instead on the thorny problem of how to land a million men, and all the supplies they would need, over the beaches.

The main attack was made by two brigades of the 2nd Canadian Division, together with the 14th Canadian Army Tank Regiment, and flanking assaults on batteries at Berneval and Varengeville were made by Nos 3 and 4 Commandos, each of which had with them a detachment from the 1st United States Rangers. The beaches of Dieppe, wreathed with barbed wire and dominated by hardened machine-gun posts, were an almost-perfect killing ground, and by the time four hours had passed, the result was in no doubt, and the

British could do nothing save try to recover as many survivors as possible into the small armada of ships which waited offshore. Blue beach, where the men of the Royal Regiment of Canada landed, claimed 483 of the 543 officers and men who waded ashore. In human terms, it was an unmitigated disaster almost all round, and of the 7,000 men committed, barely 2,500 escaped and 1,367 became prisoners of war. If successful small raids (or 'stunts', as they were known in the understated jargon of the day) were good for morale, a substantial loss like this clearly had the reverse effect.

If Dieppe had a bright side, it was the way Lord Lovat's No 4 Commando had handled the task of subduing the Hess Battery at Varengeville. Landing on beaches to either side of the battery itself, a little before 05.00, Double British Summer Time (that is, just before first light) the Commando deployed leaving Major Derek Mills-Roberts to pin down the battery with mortar fire while Lovat's larger main force worked its way round and took the objective from the rear, largely with bayonet and knife. To the cognoscenti, this limited success was only a reinforcement to the view that stealth and cunning paid greater dividends than brute force in this type of operation. And it was stealth and cunning which were the hallmark of many of the smaller-scale raids mounted at the time, by units such as March-Phillips, Hayes and Appleyard's Small Scale Raiding Force (SSRF). By and large, these raids – whose avowed purpose was simply to kill Germans in their own backyard, so to speak – were successful in that few of the members of the small raiding parties – eight or ten was normal – failed to return. There were exceptions, though, and the third mounted by the SSRF, on the Normandy coast at the eastern end of what, not two years later, was to be Omaha beach, was a complete failure.

March-Phillips' men had scored two minor successes previously, with Operation Barricade, a punitive raid on the eastern side of the Cotentin Peninsula to shoot up a gun-site, and Operation Dryad, in which they had snatched the entire crew of the German-manned lighthouse and radio station on Les Casquets, a group of rocks twelve kilometres north-west of Alderney. These were just the sort of small operations Churchill had envisaged two years before, and he lost no time in incorporating mention of their methodology into one of his speeches, saying menacingly: 'There comes out of the sea from time to time a hand of steel which plucks the German sentries from their posts with growing efficiency.'

The third foray, Operation Aquatint, was loosely enough framed, like most raids at that time, 'to test German defences, take prisoners and generally shake up the enemy'. A party of eleven raiders landed (there should have been twelve, but Appleyard had broken his ankle during Operation Dryad, and remained aboard the motor torpedo boat which ferried them across the Channel) and were detected almost immediately, by a German patrol dog (which was awarded the Iron Cross and wore it thereafter on its collar) and came under effective fire before even leaving the beach. Gus March-Phillips died straight away, alongside two of his men, while most of the rest were taken prisoner immediately. Three managed to get off the beach but were soon captured. Grahame Hayes swam three kilometres along the coast and later contacted members of the French resistance; they got him to Paris, and then to within sight of the Spanish border before he was taken; though it seems as if he was under some sort of surveillance all the time, as a result of being betrayed, even before leaving Normandy, by a traitor in the resistance cell at Lisieux, Robert Kiffer, a *résistant* who had been captured and turned by the Gestapo in 1941. Hayes was executed by firing squad at Ivry, to the south of Paris, in July of the following year, conceivably on the same day that Appleyard, the last of the SSRF's founding triumvirate, died when the aircraft he was travelling in, on a mission to seize a bridge in Sicily during Operation Husky, was shot down, probably by friendly fire.

Though doubtless downhearted at the way Operation Aquatint had gone so disastrously wrong, Appleyard was determined to salvage some sort of role for the SSRF, and the only way to do that was to bring off a successful raid. Just three weeks later he mounted Operation Basalt, but in the process the raiders stirred up a hornets' nest for themselves and others of their ilk in the shape of a vengeful directive from the OKW (*Oberkommando des Wehrmacht* – German Armed Forces High Command) that any commandos encountered from then on were to be shot, and that all British prisoners of war taken at Dieppe were to be chained as a reprisal for the binding of Germans taken captive. This time the target was occupied British territory, the island of Sark, and the English-language *Guernsey Evening Star*, under German control, of course, printed the following story about the raid and its repercussions:

In the early morning hours [of 4 October 1942] sixteen British soldiers attacked a German working party consisting of an NCO and four men. The German men, clad only in their shirts, were bound with strong cord, prevented from putting on any further garments and taken to the shore.

When the Germans offered resistance against this unheard-of treatment, the NCO and one man were killed by rifle shots and bayonet thrusts, while a further man was wounded. This fact is confirmed by the statement of a German sapper who succeeded in escaping in the scuffle. The cross-examination has brought to light that the binding had been planned by the British before-hand.[15]

The newspaper story went on to relate the countermeasures ordered by the OKW:

1. From noon on October 8th all British Officers and men taken prisoner at Dieppe will be bound. This order will remain in force until the British War Office can prove that in future only truthful declarations regarding the binding of German prisoners of war will be issued and can further prove that its order [not to tie prisoners' hands] will be carried out by British Troops.

2. In future, all territorial and sabotage parties of the British and their confederates, who do not act like soldiers but like bandits, will be treated by the German troops as such and whenever they are encountered they will be ruthlessly wiped out in action.[16]

Two weeks later, Hitler personally amplified this order with an instruction that all captured raiders should be exterminated. It was under this *Sonderbehandlung* (special handling order) that Hayes was executed, and the same treatment was meted out to, among others, five SAS soldiers captured north of Paris in 1944, and to six SBS men captured during a raid into the Aegean. Reports of the men who actually carried out the raid on Sark – they included twenty-six-year-old Appleyard and Anders Lassen, who was just twenty-two – were somewhat at odds with the German version, though that is hardly surprising.

From the outset, Lassen was clearly eager to use his fighting knife (he was no stranger to knives; during basic training he had demonstrated that by stalking and killing a stag with one, much to the delight of his comrades, who welcomed the fresh meat) and had to be restrained from tackling a sentry quite unnecessarily. Said Les Wright: 'This was my only raid with Andy Lassen, and Apple had

THE WHITES OF THEIR EYES

already stopped him from doing in a sentry who passed us as we lay doggo for a few minutes after climbing the Hog's Back.'[17] Now, as the party approached the Dixcart Hotel, where the twenty-strong German garrison on the island was billeted, Lassen took it upon himself to dispose of the sentinel there. Bombardier Redborn, another member of the party, recalled: 'Andy said he could manage the one sentry on his own. We listened to the German's footsteps and calculated how long it would take him to go back and forth. Andy crept forward alone. The sinister silence was broken by a muffled groan. We looked at each other and guessed what had happened. Then Andy came back and we knew it was all right.'[18]

The German engineers, sent to the island to build a boom defence at Creux harbour, were asleep in single rooms in an annex to the hotel, and within moments of the raiders' entry, were captives. Wright said: 'We tied their thumbs together and, after cutting their pyjama cords, made them hold their trousers up. We were just snatching prisoners for investigation. We didn't sail with handcuffs and manacles and, even after all these years, I'm still angry at suggestions that we silenced the Germans on Sark by stuffing their mouths with mud. It's like saying we were animals.'[19]

Wright could not recall gagging any of the prisoners, and Appleyard's after-action report bears him out:

> The prisoners were assembled under cover of trees near the house. In the darkness, one of them suddenly attacked his guard and then, shouting for help and trying to raise the alarm, ran off towards buildings containing a number of Germans. He was caught almost immediately but, after a scuffle, again escaped, still shouting, and was shot. Two of the other prisoners broke away, and both were shot immediately. The fourth, although still held, was accidentally shot in an attempt to silence him by striking him over the head with the butt of a revolver. [In fact, it appears that the raider in question 'A bloody big bloke – Toomai the Elephant Boy,' we called him, actually struck the man with the barrel of his pistol, not the butt, forgetting that his finger was on the trigger, 'and blew the top of the German's head off'.][20]

Redborn's memory was of Appleyard, 'shouting, "Shut the prisoners up!" This began a regular fight. My prisoner had freed his hands; I bowled him over with a rugby tackle, but he got free again. He was much bigger than me; I couldn't manage him, so I had to shoot him. Andy was still holding on to two prisoners, then lights came on in the hotel. Andy wanted to throw grenades through the windows, but Major Appleyard said: "No, we might need them later." The Germans had started to come tearing out of the hotel, so we preferred to run for it. We still had one prisoner, and he was petrified.'[21]

Wright covered the party's confused retreat with his Bren gun, and they made their way back to the beach where they had landed, scrambled back into their dory and were soon back aboard 'The Little Pisser', MTB344, an eighteen-metre boat lightened of most of its weapons and with both performance-enhanced main engines and an electric auxiliary engine for silent running, and on their way back to Weymouth. Lassen wrote in his diary, 'The hardest and most difficult job I have ever done – used my knife for the first time.' A force member who had not been on the raid, Ian Warren, was still asleep when they got back. 'Andy woke me,' he said. 'He held his unwiped knife under my nose, and said, "Look, blood!"'[22]

Knives were not the only weapons which interested Lassen as a means of silent killing; as his biographer, Mike Langley, notes:

He was keener on the bow and arrow as a raiding weapon. Reviving the bow was not an idea gained from kids' comics and adventures for boys. Lone archers had raided successfully in the Spanish Civil War – as Lassen may have heard from his father or from [Peter] Kemp [a fellow SSRF member] who had served at first in Spain with the Carlists, a royalist faction [sic] responsible for the bow's reintroduction. Carlist raiders, all in black and armed with short black bows and arrows, infiltrated Republican trenches on night raids and killed sentries silently on challenge.[23]

Lassen put his case to the War Office: 'I have considerable experience in hunting with bow and arrow. I have shot everything from sparrows to stags, and although I have never attempted to shoot a man yet it is my opinion that the result would turn out just as well as with stags.'[24]

The War Office considered his claim that a trained archer could fire up to fifteen virtually soundless shots a minute, each capable of killing 'without shock or pain'. The British War Office then arrived at a typical compromise by sending him two hunting bows with arrows – but not the permission to use them against the enemy. The bow and arrow in 1942 was classed as 'an inhuman weapon', a ruling which drew a scathing paragraph from Appleyard: 'Such is the anomaly of modern warfare that the traditional weapon of Crécy and Agincourt should be prohibited while recourse is permitted to such horrors as rockets and atomic bombs.' (Though inasmuch as Appleyard was killed long before the feasibility of the atomic bomb became known, perhaps we may be permitted to doubt the veracity of that latter attribution.)[25]

The Sark raid had very little to show by way of real results, despite some after-the-fact spin-doctoring by the War Office, who tried to imply that Appleyard's men were seeking evidence of German mistreatment of the Channel Islanders, and was by no means popular with most of the 400-odd Sarkees, many of whom had evolved a comfortable way of co-existing with the small occupying force, and who were now subjected to a much more rigorous regimen. The bitterness deepened the following year, when twenty-five islanders were deported to Germany as 'unreliable elements capable of giving information to British Commandos'[26] and set to forced labour on farms. Churchill exulted, though: 'The British Commando raids along the enormous coast,' he said, 'although so far only the forerunner of what is to come, inspire [Hitler] the author of so many crimes and miseries with a lively anxiety.'[27]

•

In North Africa, the British also looked at ways to keep the Germans off balance by raiding their rear areas. Commando units had already been posted to the Middle East. The Long Range Desert Group was formed to carry out deep reconnaissance missions behind German lines. An outfit called Popski's Private Army used locally recruited Arabs to sabotage enemy bases. The Special Boat Section was formed to conduct beach and harbour reconnaissance in canoes, and a British Army captain, David Stirling, formed another raiding group, the Special Air Service, that was to become a legend.

Destroying the German Afrika Korps' supply lines was the job of L Detachment of the Special Air Service (SAS) Brigade. This title was designed to confuse Nazi intelligence into thinking that an allied airborne brigade was in North Africa.

Desert raiding was very different from the seaborne operations used in north-west Europe. Bored 8th Army soldiers volunteered to serve in the SAS to escape the routine of conventional soldiering, and exiled French and Greek soldiers also flocked to the colours to avenge the German occupation of their homelands.

The SAS operated in small patrols of half a dozen jeeps or trucks, crammed full of supplies, water, explosives and machine-guns. To surprise the enemy they had to drive deep into the Libyan desert, travelling hundreds of miles, before swinging inland to attack unsuspecting German supply bases and airfields. Living in the desert required special training and plenty of perseverance. Desert navigation was a skill that every SAS member had to master, using the stars and sun compasses. Early SAS training was rough and ready: potential members were sent on route marches into the desert carrying huge rucksacks. Every SAS man had to be parachute trained.

The first SAS raid was carried out by parachute, with the raiding party marching on foot to its objective. Transport to the target on subsequent raids was provided by the trucks of the Long Range Desert Group. Once near their targets, the SAS men had to get past German guards, either by stealth or by silently killing them with daggers. The SAS men would then race around German airfields planting explosive charges next to Junkers bombers or Messerschmitt fighters. Setting demolition charges at night on enemy airfields was difficult and stressful work. SAS men had to work fast, setting detonators and securing explosives to targets. Even the smallest errors could mean the explosives would explode prematurely or not at all.

Next the SAS tried jeep raids. The assault force would simply drive through German airfields machine-gunning any aircraft they could find. A signal flare would start the assault, and when German aircraft were burning all over the airfield, the raid commander would send up another to order the withdrawal. SAS raiders had to have superb weapon-handling skills to get fire quickly on targets and clear stoppages. These were large operations with up to twenty jeeps attacking one airfield at a time. Once the raid was over, the SAS would scatter into the desert to confuse the pursuing Germans and then meet up at a prearranged rendezvous deep in the desert.

The purpose of Commando raids, then, was clearly to strike fear into the hearts of each and every German soldier posted to a coastal installation and supply bases deep in the North African desert. Without direct evidence, we cannot know how successful it was on a personal level, but it certainly caused staff officers some concern. Many, like Alfred Jodl, the *Wehrmacht* Operations Chief (and eventually signer of the instrument of surrender), believed the commandos to have been culled from the ranks of violent criminals in the nation's prisons: 'The fact that many previously convicted persons and criminals were included in the Commandos, who were of course reckless people, was proved by the testimony of prisoners.' (And one may be forgiven for wondering how this differed, in spirit, at least, from the German habit of drafting men from the punishment battalions to undertake hazardous duties. Compare, perhaps, the recruitment of *Flammenwerfer* operators on the Eastern Front.) The British official position was contradictory, of course: 'We never enlisted anyone who looked like the tough-guy criminal type as I considered this sort would be cowards in battle,' said Brigadier John Durnford-Slater, rather disingenuously, when discussing selection criteria for No 3 Commando.

In fact, even such luminaries as Robert Blair 'Paddy' Mayne were recruited from guard rooms (Mayne was awaiting court martial for laying out his commanding officer and then chasing him out of the Mess with a bayonet. He finished the war in command of 1 SAS, a Lieutenant-Colonel with a DSO and three bars), if not actually from prisons, and there are long-standing rumours that Patton's Third Army included one unit released from custody to train for a 'suicide' mission. 'They all came from death row, were dirty, nasty and constantly fought amongst themselves,' swore a US Army cameraman who said he came across the group by accident, somewhere in southern England. 'One was a Navaho Indian, another was black and others came from various ethnic backgrounds in urban cities. Some were dangerous psychotics charged with such offences as murder and rape who had been given the option of signing up for a suicide mission or facing execution.' Needless to say, the US Army has no record of such a unit ever existing. No doubt the cameraman in question was almost tempted to turn the story into the script for a film, perhaps with a title something like *The Nasty Nine*, *The Terrible Ten* or *The Evil Eleven....* But we digress.

Not every commando raid was an exercise in pure terrorism; some, like that directed at the radar station at Bruneval, carried out by men of the 2nd Battalion, Parachute Regiment, under the command of then Major John Frost, who was later to lead them into battle at Arnhem, had as its object the discovery of just how far German radar technology had advanced. Another raid, on a hydro-electric plant at Glomfjord in Norway, carried out by men of No 2

Commando, was aimed at disrupting production at a nearby aluminium smelting plant. Some raids had a tactical purpose within a larger battle plan, too, such as that on Walcheren Island at the mouth of the Scheldt, which dominated the approaches to Antwerp, itself taken intact just prior to the raid by the British 11th Armoured Division. Three Royal Marine Commandos and No 4 Army Commando assaulted the island, neutralizing the defences and taking the town of Vlissingen (Flushing) despite suffering heavy casualties.

Some of Lassen's friends dated a very obvious blood-lust in him to the Sark raid, and his first experience of killing a man with a knife. That there was such a lust in him is in little doubt, as his later exploits in the Mediterranean show. A comrade-in-arms there, Adrian Seligman, later described him, in a script for a radio programme: 'He was brave with a calm, deadly, almost horrifying courage, bred of a berserk hatred of the Germans who had overrun his country. He was a killer, too, cold and ruthless – silently with a knife or at point-blank range with pistol or rifle. On such occasions there was a froth of bubbles round his lips and his eyes went dead as stones . . .'[28] In this, he was out of the ordinary, even within the limited concept of the word as used of special forces soldiers, but none the less, we can continue to follow his career through to his death in 1944, and be quite sure that it will lead us into close-quarter battle at every opportunity.

Lassen left the SSRF when it became clear that a change of role was soon to be forthcoming, and was transferred (appropriately enough, for he had been a merchant seaman before volunteering for special forces) to the fledgling Special Boat Squadron in Cairo in February 1943. The SBS had been formed, by George [now Earl] Jellicoe – son of the commander of the British Fleet at Jutland in 1916 – after the division of David Stirling's Special Air Service into two, the other part becoming the Special Raiding Squadron, under Paddy Mayne (Stirling himself now being in German captivity). The new force also included fifty-five members of the old Special Boat Section, and remnants of the SSRF, including Lassen. He was soon in action against Kastelli Pediada airfield near Heraklion on Crete's north coast, in an operation very much like those conceived and executed by Stirling and Mayne in North Africa. But where the original SAS raids had used vehicles, Lassen and his men had to make their way on foot some hundred kilometres across the mountainous spine of the island before sighting their target.

Eight Ju 87s and five Ju 88s, together with a handful of fighter aircraft, were dispersed around the airfield when the party arrived in position. They split into two pairs, and Lassen himself chose to deal with the Italian sentries facing him and his companion, Corporal Ray Jones: one with his fighting knife, in the approved Fairbairn and Sykes manner, hooking his free hand around the man's jaw and wrenching his head around to expose the entry point for an attack on the subclavian artery, the other by shooting him without bothering to remove his pistol from his pocket.

This shot roused the guard, and Lassen and Jones had to run for it, making their way back half an hour later when the situation had calmed down somewhat. For his part in the operation, Lassen won a bar to his Military Cross (originally awarded after the Sark raid), while the other members of the party were awarded Military Medals, being 'other ranks'. The citation for Lassen's award talks of how he returned to the airfield to finish the job and was then obliged to shoot his way out of trouble, thus creating a diversion which allowed the other two-man team to destroy three aircraft and a fuel tank. 'It was entirely due to this officer's diversion', the citation reads, 'that planes and petrol were successfully destroyed on the eastern side of the airfield since he drew off all the guards from that area.' Sergeant Jack Nicholson, who was paired with Corporal Sidney Greaves to make up that other two-man party, rejected the official view completely. 'Andy Lassen was a brave man and a cool man,' he told Lassen's biographer. 'I would call him stupidly brave, except that he kept on getting away with it – and sometimes he got credit for things he didn't do. Kastelli airport is an example. That was no diversion, that was a bungle. Lassen and Jones were supposed to be as silent as me and Greaves.'[29]

After a spell with an irregular sub-unit known as the Levant Schooner Flotilla, which operated a collection of decidedly (and deliberately) scruffy caiques out of Beirut to attack targets in the Aegean, Lassen returned to the larger SBS fold with a raid, which on 17 September 1943 'captured' the island of Simi in the Dodecanese from the Italians who had garrisoned it, just at the time of the Italian capitulation. He was to play a major part in the defence of the island when a party of 120 German paratroopers arrived shortly after, and won a second bar to his MC there after killing three more Germans with his fighting knife despite having badly burned both his legs. Hank Hancock, an SBS sniper who went on the Simi operation, said of him: 'Once he got going, he'd kill anybody. He was frightening in this way – and his view of the Germans was more personal than ours because his country was occupied. I think he was driven by the occupation although it's difficult to assess if the killing instinct used the war as an excuse or whether it sprang from genuine hatred of the enemy. But I do know that if he had the opportunity he'd kill someone with a knife rather than shoot, which seemed a bit odd to me.'[30]

And then, after a period during which he was hospitalized for illness, rather than wounds, a number of small 'butcher-and-bolt' terror operations and another spell in hospital after being accidentally shot by a member of his own patrol while hunting Germans on the island of Calchi, came 'Andy Lassen's Bloodbath', the raid on Santorini. Santorini, or Thera, to give the island its ancient name, is perhaps the most remarkable in the whole Mediterranean. Sometime around 1450 BC – the exact date is unclear – the entire volcanic island, then roughly circular in form and some 32 kilometres in diameter, blew up, destroying the Classical Minoan civilization on Crete, 96 kilometres away, in the process, and leaving an incomplete ring of land surrounding a flooded caldera 22 kilometres across. It is said that the eruption was accompanied by earthquakes which caused tidal waves in the Red Sea, giving rise to the biblical account of the waters parting to allow Moses to lead the Israelites out of captivity in Egypt.

Be that as it may, the island in modern times is a very strange place indeed, with its one major town, also called Thera, perched at the top of sheer cliffs hundreds of metres high overlooking the central crater-cum-lagoon. In April 1944 it was garrisoned by little more than a platoon of Axis soldiers, reportedly thirty-eight Italians (for not all Italians had subscribed to the separate peace) and ten Germans, billeted on the first floor of the Bank of Athens in Thera but also manning a radio station some distance away. 'Operation No 13' had, as its object, the destruction of enemy shipping, the destruction of enemy communications and personnel and to attack any other targets of opportunity which might present themselves. It was undertaken by a party of twenty men, including a Greek interpreter, nominally divided into two ten-man patrols, who landed on the eastern shore of the island, on a beach composed of black volcanic sand, on the night of 22 April, after a three-day journey from a staging post in Turkey.

Surprisingly, for an operation commanded by Lassen, who had no time for written analyses, there exists a detailed after-action report, presumably, by its phraseology, actually written by someone else, though it is signed by him (Lassen's usual reports, his biographer tells us, were more likely to be extremely terse. 'Landed, Killed Germans. Fucked off,' was the way he reported one raid, his men maintained). It does not differ in great part from the recollections of one member of the raiding party, Corporal Sammy Trafford, whom we last encountered describing the goings-on at the Killer School in Jerusalem:

It was my first raid with Lassen – 'Andy Lassen's Bloodbath' they called it afterwards because he killed nearly every bugger. [Jack] Nicholson said to me as we were setting off: 'Coming with the killers, are you, Sammy? Don't worry, you'll be all right.'

Lassen halted us on the march to the town and made everyone swallow two tablets of benzedrine, watching them go down before taking two himself. He wanted us wide awake. He was a good organizer, a hitman and killer. He carried just his Luger pistol and a fighting knife, and it was said that he was a devil with the knife.

Fear didn't seem to bother him, although I suppose he felt some; I tried to imitate his fearlessness and easygoing ways in the face of danger. He walked on the balls of his feet, very quietly. Someone once described him silently framing himself in a doorway; Andy Lassen never stood in a doorway in his life. He'd been taught better. One moment he'd be outside, the next inside and beside us . . .

Sean O'Reilly and myself were pushed first into the bank by Lassen. Five doors faced us. I took the middle door and O'Reilly took the one on the right while Lassen brought in the others, one at a time. O'Reilly shouted 'Grenade gone' to let everyone know he'd thrown one into his room, but then his Schmeisser jammed and he began cursing: 'Jesus Christ, where the bloody hell are the others?'

Suddenly [Lieutenant Stefan] Casulli was hit by fire from a door that I'd not seen. He staggered towards me, obviously dying. I was hit next from the same direction. The wounds were in my upper arm and left leg. I staggered, too. Next Guardsman Jack Harris shouted: 'I've been hit in the leg,' whereupon he limped out on the terrace. Sgt Kingston from the RAMC should not have been on the raid but had asked Lassen to take him. Lassen agreed, but told him not to go inside the bank because he couldn't

COMMANDOS AND SPECIAL FORCES

be both a killer and our medical orderly. Kingston died from shots in the stomach.[31]

Signaller Billy Reeves, who had been to war with Lassen before, takes up the story:

Casulli and Kingston were exceptions to the rule that very few people got killed on Lassen's raids. He was a lucky officer. But the bank was different. There was only one outside door – where I got knocked to the floor – and the enemy were on the upper floor so we couldn't go in saying 'hands up'. I remember us shooting one of them who tried to escape through a back window. Lassen's motto on prisoners that night seemed to be: 'Don't take any.'[32]

The after-action report goes into a little more detail:

On the first floor of the Bank of Athens at Thera was a reported billet for 38 Italians and 10 Germans. This report was partly false; there were less than 35 men in the billet. We succeeded in getting the main force into the billet unobserved, in spite of barking dogs and sentries. The living quarters comprised twelve rooms.

It was our intention to take the troops there prisoner. This idea had to be abandoned and will have to be abandoned in similar circumstances in the future until raiding parties are issued with good torches. Casualties were sustained during the general mix-up in the dark. Instead [presumably, of taking prisoners], the doors of the rooms were kicked in, a grenade thrown into the room and two to three magazines of TSMG [Thompson sub-machine-gun] and Bren emptied into each room.

At approximately 02.45 hours, and when I was satisfied that all the enemy were killed or wounded, we left the building carrying Sgt Kingston. Shots were exchanged with stray enemy during this withdrawal. We returned to the cave to find Lt [Keith] Balsillie [who had been detailed to attack the wireless station at Murivigli, with four ORs] waiting for us with eight prisoners.[33]

Not everyone shared Lassen's obvious enthusiasm at the whole-sale destruction. Sergeant John Nicholson: 'That was the only time

I was in action side-by-side with Lassen and it's one of the reasons why I'm trying to forget the war. It's no fun throwing grenades into rooms and shooting sleeping men.'[34]

After describing the action, Lassen went on to make some general observations about tactics and training which bear out Sammy Trafford's: 'There is no doubt that shooting up barracks at night requires a great deal of skill and experience, such as only the older men in the SBS have and which will not be found in reinforcements. Lack of experience must be made up by rigid training, especially in street and house fighting, and they should be generally taught how to look after themselves, not to stand in front of doors for example.'[35]

After Santorini, Lassen's field of activity widened to include the Adriatic as well as the Aegean, and that meant launching raids into Yugoslavia and Albania. Here there were much better organized partisan activities, and Lassen came into contact with people no less bloodthirsty – and probably considerably more cruel – than he was himself, such as the Yugoslav partisans who, according to a hearsay account retold by Billy Reeves, extracted a terrible revenge for the rape of one of their comrades by four German soldiers: 'We were told they were staked out naked on the ground while a girl moved among them, gently masturbating each man. Behind her came the executioner, with a razor-sharp knife. The penises were stuck in their mouths.'[36]

In fact, Yugoslavia was in the grip of a war within a war, between the only partisan band the Nazis nurtured, the Ustashi, and the Allied-sponsored Chetniks. The situation was actually much more complex than that, with at least half a dozen factions seeking to profit from the situation in one way or another, all with a history and tradition of extreme violence and none with any regard for any save its own ethnic, religious or political grouping.

Anders Lassen died in action, on 9 April 1945, less than a month before the war in Europe was to finish, aged twenty-four. He was involved in what could be properly called a battle for the first time, attempting to dislodge around 1,200 Axis troops, many of them nomadic Muslims from Turkmenistan, from the town of Commachio and the adjoining harbour of Porto Garibaldi, to the south of Venice. And uncharacteristically, he died while trying to take prisoners, but with no less aggressive energy than he had been demonstrating since first going into action, three years earlier:

'All our [earlier] success had come from stealth,' said his Sergeant Major at Commachio, Les Stephenson:

but now we were being used for an infantry assault that, considering our type of soldiering, was a suicide mission. We were rushing along the road [actually a causeway] with him [Lassen] in front, blowing his whistle and shouting: 'Come on. Forward, you bastards. Get on, get forward.'

I caught up with him at a place where the machine-gun fire was getting heavier and pinning us down. Then it eased off for a while; the defenders would have been as much in the dark as we were. We had taken two of their positions and they were not to know we were only seventeen men. For all they knew, we could have been a couple of hundred.

Lassen was running out of grenades when I joined him in attacking the third German position. He was a grenade man . . . very fond of grenades. Some people hardly used them but he had a good throw and could put a grenade where he wanted. He said: 'Have you any grenades, Steve?' so I passed him mine which I'd been keeping in reserve because we were supposed to go further along. He said: 'You bastard, you haven't pulled the pin out!' You don't run around with a loose pin, so I let that remark pass over my head . . . He snatched at least one grenade out of my hand before I could pass it to him.

He shouted to the pill-box in German to surrender. Although a ruthless man, he wasn't brutal and, on occasion, would offer an opportunity to surrender. Somebody shouted 'Kamerad', so he stood up from behind the small rise in the road that we had been using for cover. He told the rest of us to stay put while he went across in the darkness; as he neared the pill-box, there was a burst of machine-gun fire and then silence. We could barely see the pill-box, let alone what was going on. The silence seemed to last for twenty minutes but could only have been a few seconds and then I heard him call . . . 'SBS, SBS . . . Major Lassen wounded . . . Here.'[37]

Lassen was shot in the abdomen or lower chest, and died very soon after. Stephenson goes on:

His decision to capture, not kill, had been the undoing of the whole thing. But I don't think he was killed by treachery. I think it was just bad luck. They had shouted 'Kamerad' and were going to come out and give themselves up when they saw a figure rushing towards them in the darkness with a weapon in his hand. That could have alarmed them. Perhaps he shouldn't have gone forward so quickly to accept the surrender, but he was quick by nature and action. He didn't lead from behind; he was always in front. Besides, it wasn't an occasion for pussyfooting around. He wanted the pill-box silenced quickly, and the initiative kept going.[38]

Lassen was awarded the VC for his final, fatal assault. The two corroboratory personal reports required to procure such an award came from Bdr. T.C. Crotty and CSM C. Workman. Colonel David Sutherland, Lassen's long-time direct superior, who was by then commanding the SBS, put his name forward for the premier award since it was the only decoration which could then be awarded posthumously, despite his colleagues advising him against the move. He was successful.

Of course, Anders Lassen's exploits formed only a small – though exemplary, in all senses of the word – part of SBS's activities in the Mediterranean, even if some of his men did feel that the Dane had volunteered them for all 381 operations the squadron actually carried out. In Italy and in the Balkans, a variety of Commando units, almost all of them known by a virtually indecipherable string of initials, went into action in enemy-occupied territory, not just raiding but setting up clandestine operational bases occupied over longer periods. This type of activity reached its peak in France, before and after Operation Overlord, with the insertion by parachute of teams of heavily armed and equipped SAS soldiers, sometimes in as much as company strength, into sites such as 'Houndsworth', south of Dijon, and 'Kipling', near Auxerre, to disrupt German communications and logistics.

Some of these operational bases were substantial – Houndsworth, for example, comprised a total of 153 officers and men, and had fourteen jeeps armed with twin Vickers Type K machine-guns and even a six-pounder anti-tank gun. The SAS soldiers were quite ready to fight it out, if necessary, with any German patrol they encountered, and Houndsworth alone is credited with accounting for an estimated 220 German dead at a cost of one officer killed and four troopers wounded.

The SAS detachments often worked in tandem with French *maquisards*, many of whom were ex-soldiers, and in general, relations between the SAS and the French were rather less strained than they had been with the unpredictable Yugoslavs, but – especially after D-Day – there was increasing friction between Communist and Gaullist groups, who were bitterly opposed to each other and not at all inclined to co-operate, conscious that the day was fast approaching when a civilian government would be restored. In some cases, however, SAS operations were compromised. 'Bulbasket', for example, a base south of Châteauroux, was betrayed by unknown local elements on 7 July; the base came under attack and all thirty-seven men there died, some almost certainly executed after they had surrendered, according to the provisions of the *Sonderbehandlungs* order, of which they were ignorant.

Many of these *maquisard* bands were supplied from London via 'Jedburgh' teams of Office of Strategic Services' (OSS) and Special Operations Executive (SOE) operatives, usually one American, one Briton and one French, not all of them men, by any means. In fact, women were regarded by many as being much better fitted to these clandestine operations; many were recruited and not a few gave their lives, some at the hands of the Gestapo's torturers. Resistance to the Germans was most open in the Balkans, and by 1942 Yugoslavia was engulfed by war as Partisan groups began to inflict heavy losses on the Germans.

Tito's Partisans were the most successful exponents of guerrilla warfare during the Second World War, keeping more than twenty Axis divisions at bay for three years before taking the offensive and driving them from Yugoslavia. The Nazi occupation of Yugoslavia was brutal, and within weeks of the fall of the country to Hitler's panzer divisions, ordinary Yugoslav citizens were taking to the mountains to begin the resistance against the hated Germans. Already in the mountains were Communists loyal to Josip Broz – Tito – who began forming Partisan main force brigades. The Partisans formed three types of units: the main force brigades were highly mobile light infantry designed for offensive operations; local militia units who only operated in a specific area on static guard work; and espionage and sabotage networks, who operated in civilian clothes in areas occupied by the Germans and their allies to collect intelligence and kill individual Nazi soldiers.

Until the British SOE started supplying guns and ammunition in 1943, the Partisans had to rely on old Royal Yugoslav Army weapons or captured German equipment. The German MG42 was a favourite weapon because of its firepower, and this weapon was later adopted by the postwar Yugoslav Army as the M53. For close-quarter battle, captured MP40 Schmeissers or British Sten guns were favoured. Partisan training was rudimentary and brief. Basic weapons handling was given to those who did not come from peasant stock and had not been brought up to hunt and shoot.

The Partisans could not at first hope to take on and defeat the Germans in open battle, so they concentrated on hit-and-run attacks against Nazi communications and garrisons. Small German outposts guarding bridges or roads were ambushed and destroyed. Attacks on large German garrisons were more sophisticated. A Partisan main force brigade would first take up positions on the high ground surrounding the enemy-held town. Then a handful of Partisans would infiltrate the town in civilian clothes at night to occupy important buildings. They would then open fire on the Germans and force them to counter-attack. Teenage and female Partisans would also be used to drop grenades through the windows of German barracks and offices. At this point the main Partisan force would attack from outside the town, capitalizing on the confusion in the enemy ranks.

Escaping from the inevitable German pursuit was more difficult. The Germans regularly mounted massive cordon and search operations to trap the Partisans' main force brigades in their hide-outs. The only way to escape was to march over the mountains to safer areas. These retreats were brutal, no-holds-barred affairs. Only weapons and equipment that could be manhandled up a mountain track could be taken. Wounded and sick who could not keep up were shot rather than let them fall into German hands. Every so often a small rearguard would be left to delay the German pursuit to give the retreating column time to escape. They would set up a road block and then place mines around it. A few machine-guns on high ground would kill any Germans who tried to tackle the obstacle.

There were more or less effective resistance groups in all the countries overrun by Germany, many of them led by army officers who had managed to escape the round-up of dangerous elements which always followed. Others, who managed to escape to Britain, were trained by the SOE, and its American equivalent the OSS, and returned to their homelands both to act as cadres and to carry out specific missions themselves. By definition, all these missions come under the heading of close-quarter battle, and were carried out with knives, small arms and grenades. Perhaps the most important was the assassination of Reinhard Heydrich, Protector (i.e., *de facto* ruler) of Bohemia and Moravia, in a daring attack carried out by SOE-trained Czechs on 27 May 1942. Heydrich died of his wounds on 4 June, and as a result the village of Lidice, which had a very tenuous connection with the resistance (two families had sons serving with the Free Czech Forces in Britain), was systematically destroyed (even the graveyard was excavated), all its adult males shot, its adult females sent to Ravensbruck concentration camp and its children dispersed. Revenge of this fashion was a regular part of the Nazi struggle to control Partisan activity but often had just the reverse effect to that intended, ensuring even greater hostility and will to resist in the population. Heydrich's assassins were betrayed and cornered in their hiding-place, the Karel Boromejsky church in Prague, where they committed suicide after a brief gun battle.

There was also at least one plan to kidnap or assassinate Adolf Hitler – Operation Iron Cross – prepared by the OSS but never actually mounted. Captain Aaron Bank, who had operated with a Jedburgh team in Provence as part of the lead-up to Operation Dragoon, was to have parachuted into Austria with a guerrilla force recruited from disenchanted German PoWs during the winter of 1944, and infiltrated Hitler's alpine redoubt, where the Führer's last stand was to be played out with a supporting cast of thousands of fanatical SS men. Bad weather prevented the operation getting under way in time, and in any event, the last stand came to naught, for Hitler committed suicide in his Berlin bunker rather than face the wrath of the Russians.

Most assassinations, needless to say, had rather more mundane targets – minor German functionaries, particularly efficient (or brutal) SS officers or, very often, traitors. The acts themselves were hardly more than sordid little murders, though with an importance somewhat inflated by their purpose:

Bartek must be out of his mind to expect me to do the job in such conditions! With my .38 Colt in my overcoat pocket, I pulled on an armband inscribed *Im Dienst der Deutschen Wehrmacht* ['On German Army Service'], which was necessary for movement in the streets after police hours, and slipped out of the house.

As I walked up and down at the tram-stop I kept watching the entrance to the Gastronomia. The movement helped clear my thoughts a bit. Somewhere on the first floor, that creature Ganther [a would-be infiltrator and Gestapo informer] sat watching floor shows and listening to music, at the same time rendering a report of his infamous activities. He kept me waiting a long time . . . My nervousness increased every minute. Around midnight a shining limousine drew up in front of the entrance to the Gastronomia, and five minutes later, Ganther emerged, accompanied by two senior Gestapo officers . . . There was protracted leave-taking . . . [before] the officers got into their car, and Ganther started walking towards the Central Station.

I decided to bump accidentally into Ganther. To do that I had to outdistance him. I crossed on to the other side of the street and, walking at a quick pace, passed him; then, crossing back to his side, met him head-on. I greeted him very cordially.

Ganther never batted an eyelid. His simulated welcome was as cordial and as enthusiastic as mine. I said there were too many Germans about [to talk safely] and suggested walking along towards the Poniatowski Bridge. He agreed without great conviction, but I don't think he suspected anything. Probably he took me for a not-too-bright brawler with an incidental taste for sabotage. [We reached a point] about one hundred and fifty yards down the Aleja Maja and stopped, and at that moment I fired at point-blank range. He fell like a log. As I bent over him to take all his papers and documents, he gave an ear-splitting scream of 'Polizei!' I managed to finish him off with a bullet in the head before being fired on by the nearest police patrol. I didn't lose my head. I threw my pistol far out onto the lawn and hung my light-coloured raincoat on a small tree, and only then started sprinting along a hedge towards Smolna Street. The patrol must have seen me, for bullets were splashing all over the pavement and ricocheting. Before I had run more than twenty-five yards I was brought down by a bullet which caught me above my left ankle.[39]

Against all the odds, 'Mira' – who had previously tried to assassinate another informant named Koch, but through a combination of nerves and inexperience had only succeeded in wounding him – survived brutal interrogation at the Gestapo's Aleja Szucha headquarters, and then imprisonment in, successively, Pawiak Prison, Auschwitz, Nordhausen and Bergen-Belsen.

The Allied invasion of Nazi Occupied Europe in June 1944 was only successful because of the efforts of British and American paratroopers who landed in Europe ahead of the main invasion force. In the early hours of D-Day, small groups of paratroopers seized bridges, gun batteries, enemy headquarters and other key points. These attacks threw the Germans off balance, and it took them several days to realize the enormity of the threat they were facing from the Allied bridgehead in Normandy. The paratroopers fought in isolated groups for several days against overwhelming odds, until the main Allied armoured forces could come ashore and press inland. Those days made the difference to the success or failure of D-Day.

Two US airborne divisions, the 82nd 'All American' and 101st 'Screaming Eagles', were part of the first wave of Allied forces to land in Normandy on D-Day. They had spent two years preparing for the jump which was to spearhead the liberation of Europe.

The paratroopers who made up these divisions had almost all joined the Airborne straight from draft induction centres in the United States. In two years they all qualified as parachutists and then carried out intensive training to mould them into fighting units. Late in 1943, the Airborne moved to England to begin large-scale rehearsals for their role in Operation Overlord. Every platoon received detailed briefings on their role in the invasion, and they were shown aerial reconnaissance photographs and scale models of their objectives so each man knew his part in the operation.

This level of preparation was necessary because once the parachute drops began, the organization and structure of the Airborne divisions, regiments, battalions and companies would break down as paratroopers and gliders were blown off course by winds or were dropped in the wrong place by the US Army Air Force. Units were scattered over hundreds of square miles, so out of 6,600 paratroops of the 101st Division dropped on D-Day, only 1,000 made it to their objectives.

Once on the ground the paratroopers were on their own. They had only those weapons and equipment they landed with. The standard weapon of the US Airborne infantry was the M1 Garand automatic rifle, but many paratroopers discarded it in favour of the lighter and more compact M1 Carbine. Its 300-metre range was ideal for close-quarter combat, and it was more easily strapped on to the body for parachuting.

After landing, the paratroopers had to make contact with their comrades and begin to fight the Germans. This was a nerve-racking business because the paratroopers could not be sure if it was a German or American on the other side of the hedge. To help identification at night, each paratrooper was issued with a cricket clicker, which proved invaluable for distinguishing between friend and foe. Groups of paratroopers started to band together, getting bigger and bigger as they moved towards their objectives. Here the preparation and planning started to pay off. Every paratrooper knew his objective and, unless injured, started to move towards his objective independently, meeting up with comrades on the way. The highest ranking officer or NCO took charge of these ad hoc groupings, to achieve the objective. Sometimes generals commanded squads made up of staff officers, or sergeants led hundreds of men when no officers could be found. The paratroopers then fought for up to two weeks until relieved by heavy mechanized units. They had to improvise their defence using whatever weapons were at hand, including scores of captured German machine-guns and mortars.

It is impossible to estimate the real effect special operations had during the Second World War. They eventually occupied the attention of 50,000 Allied servicemen and women, certainly, but there can never be any estimate of the real cost of the destruction they wrought, the number of enemy troops they tied up, or even of how many deaths they were responsible for. In the wider view, they represented only a tiny proportion of Allied war effort, and it is perhaps significant that the Germans never set up even one full-time special forces organization – the raid led by Otto Skorzeny to free Mussolini from the castle on Monte Grappa where he was held after the Italian separate armistice was signed was the only substantial example of a German commando operation, rumours about raids launched on England notwithstanding, and that was carried out by regular paratroopers.

What the special forces had established was a reputation for carrying out difficult assignments against the odds with a minimum of fuss and, because the responsibility for their acts was ultimately at least partly deniable, a minimum of risk to any but themselves should they be caught in the act. Those units – like the US Rangers, The Royal Marine Commandos and the Parachute Regiment – which had established for themselves a place in the regular order of battle, kept that place after the war was over, but the most unconventional, such as the Special Air Service Regiment, were soon disbanded, only to be hurriedly raised again a decade later, when Britain found herself enmeshed in a completely unforeseen set of difficult circumstances during the break-up of her erstwhile empire.

Chapter Five

Britain's Brush-Fire Wars: the end of Empire

In the period immediately after the Second World War, the map of Europe was redrawn – or rather, recoloured – once more, just as it had been in 1918, with a wide swathe of territory encompassing six nations, stretching from the Baltic almost as far west as Denmark, to the Black Sea as far south as Turkey, now controlled from Moscow. Thus was a new empire being created even as the old ones were being swept away. It proved less enduring than the older models – it lasted less than fifty years – and much to everyone's amazement, crumbled easily into impotency rather than resisting to the bitter end.

Outside Europe, the biggest and farthest-reaching changes came in what had been the British Empire. First came the withdrawal from a mandated territory which had long proved difficult to govern, and soon became impossible – Palestine – which was to be partitioned by United Nations (UN) decision into two, a homeland for the Jews and room for the other indigenes, the Arabs. The newly formed UN studiously ignored reality – an element which the Jews, in particular, had learned over the previous decade to respect above all others, and stood by almost helplessly when a fanatical civil war broke out as both sides fought to become the *de facto* occupiers of as much territory as possible. Even while they were quitting Palestine, the British were also in the process of a much more emotionally loaded parting, from India, which was also to be partitioned into two separate states – India itself, and Pakistan, itself a curious nation in two halves, one east of Calcutta, the other in the west, hard against the borders with Iran and Afghanistan and extending up into the Northwest Frontier, towards China. The division of India was along purely religious grounds, and saw communal violence on a vast scale between the Sikhs and Hindus on one side and the Muslims on the other – some authorities believe up to a million people died: 'Every hundred yards down the track of the railway journey of some hundreds of miles [from Peshawar in Pakistan to Allahabad, in India],' recalled Edward Gopsill, a British officer serving with the Gurkhas:

> there were never less than two bodies in sight. We pulled into Amballa station and there I noticed a train standing in the siding full of dead people. Every one of them had been slaughtered. The Colonel ordered me with my company out on to the track where there were thousands of Sikhs, who had just perpetrated this carnage. So I told the Sikh leader – I said, 'You will clear this railway station. And he told me in Urdu that I could go to Hell. I thereupon lined my Gurkhas up, told them to fix swords [i.e., bayonets], and then I said, 'Unless you do clear the station I will clear you at the point of the bayonet. They held their ground for a moment until I gave the order: 'Two paces interval advance. Down came the bayonets into the on-guard position and the men started to move forward . . .'[1]

We shall encounter Edward Gopsill again soon, for he transferred the same year, if not out of the frying pan into the fire, at least to an adjoining pan.

The British left both India and Palestine just in time to turn their attention to a new conflagration; a Communist-inspired plot to drive out the British imperialists and draw resource-rich Malaya into the orbit of Beijing was gathering momentum, and was soon to erupt into guerrilla warfare. Within two years, the death toll had reached over a thousand on each side, and the Malayan Races' Liberation Army (MRLA), as the insurgents styled themselves, were clearly winning a war of attrition aimed at police posts and European-run plantations such as that of Peter Lucy and his wife Jenny, just eleven kilometres from Kuala Lumpur.

During one two-week period in 1951, the Lucys were attacked no less than twenty-five times. Their defensive policy did not vary: Peter and two 'special constables' (trusted members of his workforce who, having been sworn in, could then be armed) took up defensive positions around the house, armed with Bren guns, while Jenny manned her light machine-gun from the doorway of her children's bedroom. As an eminent British infantryman, Michael Dewar, noted in his book on the British Army's minor campaigns since 1945: 'The remarkable sight of this extremely beautiful and elegant woman prone behind a Bren gun caught the imagination of everyone ... and typified the determination of the planters not to be intimidated.'[2]

Over the decade which followed, the British in Malaya 'wrote the book' on counter-insurgency operations in jungle terrain – and also learned much that was applicable to the same sort of warfare in other places, but that is not to say they didn't make mistakes along the way. To quote Dewar again:

> It was the infantryman with his rifle on patrol that accounted for the vast majority of enemy kills. Although heavy bombers, artillery and even Royal Navy ships were used to pound the jungle with high explosive these means of mass destruction were largely ineffectual ... Massive concentrations of troops sometimes produced a miserable return. One such operation was mounted in the Ipoh area between July and November 1954. It involved RAF bombers, 22 SAS [Regiment] and four infantry battalions [that is, a total of over three thousand men]. Only fifteen rebels were killed.[3]

The seizure of power by Mao Tse-Tung's Communists in China led to other Communist and revolutionary groups beginning guerrilla campaigns to overthrow the remaining European colonial powers in the Far East. In British-ruled Malaya, the Communists struck in 1948. Ironically, the Malay Communist Party (MCP) had been Britain's ally during the war, and it still possessed thousands of weapons supplied by the British to help it fight the Japanese.

The MCP operated from bases deep in the heart of the Malay jungle, which it had been preparing ever since the British returned to Malaya in 1945 after the Japanese surrender. Its guerrillas were mainly of Chinese ethnic origin, and had learned their trade fighting the Japanese. Favourite weapons were British Sten sub-machine-guns, Bren light machine-guns, SMLE rifles and hand grenades. The geographical remoteness of Malaya meant the guerrillas received few Soviet or Chinese weapons, so the Malay Communists did not fight with the famous AK-47, the revolutionary's favourite small arm.

To force the Malay people to turn against British rule, the Communists launched a campaign of assassination and ambush against police and anyone suspected of supporting the colonial regime. Pairs of Communists would dress in civilian clothes to carry out assassinations of European farmers, policemen and their families, government officials and off-duty soldiers. Single gun shots and booby-trapped grenades were the favourite tactics. Squads of Communists would stage platoon-sized assaults on local police posts, using old British Bren guns to kill sentries, and then guerrillas would throw grenades inside to kill the occupants.

Deep in the jungle, the Communists played a game of cat and mouse with British Army special forces patrols which were trying to find their base camps. They would mount ambushes to catch the British unawares. However, the Communists realized that they did not have the firepower to stand up to the British Army, so they quickly withdrew deeper into the jungle as soon as a sizeable British force got near. To survive in the jungle, the Communists were dependent on food and medicine supplies from sympathizers in towns. Without anti-malaria drugs, though, they were vulnerable to the debilitating tropical disease. Twelve years on the Communists were defeated, and the last remaining group of them crossed the border into Thailand to live a life of exile.

The Gurkhas, who were for many years the mainstay of the British infantry force in Malaya, had one huge advantage over the young conscripts of the county regiments: they were professional fighters whose background was not entirely dissimilar from that of the men they were fighting. But that didn't mean they were invariably lucky, especially at the start, when the full dimension of the MRLA was still unclear. Edward Gopsill, now commanding a company of the 1/6th Gurkhas, was sent to investigate reports that another company had fared badly in an MRLA ambush near the Thai border:

This was very rough, hilly jungle and the going was very hard with full packs. However, we made good going and got there at about two o'clock in the afternoon. It was a scene of utter destruction. [The company commander] Ronnie Barnes's body was on the path and Tulprasad [a Gurkha officer] had got up almost to the enemy but the enemy had dug in along a ridge and there was no way that any man going up that ridge would have lived and certainly there were many bodies there. We then started the awful task of carrying back the bodies with about ninety men. I know I was carrying three packs and two Bren guns so as to release two men.[4]

Gopsill's company bivouacked for the night, and next morning set off in search of the ambush party. They found them largely by superior fieldcraft, encamped beside a tributary to the Badak river:

As we were going up, very cautiously, as you may imagine, the firing started at us and came zipping across over our heads. We couldn't see anything at that stage but we started to move forward, firing as we went. We came across this little sentry post, which by then had been abandoned, and then quite a broad track going up into a huge camp for about a hundred and fifty men. It actually had a basketball ground in it and this was where they'd made their base camp, way out of the area where they'd done the ambush on A Company. We only got eight of them, but at least we got back Tulprasad's map case and Ronnie Barnes's shoulder tabs. And then we harried them for five or six weeks all along the border and I suppose we took a toll of very nearly thirty-five of them in little actions, because they split up into smaller parties. On one of them I remember we were coming down off a spur and Partap [Gopsill's Gurkha officer] was just behind me and suddenly he pointed – there was a mark, wetness on this tree trunk. I said 'Rain', but he pointed and there was the odd bubble on the ground. Someone had peed there – and then up on the breeze came the distinctive smell of a Chinese cigarette. I had six men with me and we crawled down and there were four of them sitting around – and, of course, they were despatched straight away.[5]

Newly arrived short-service conscripts suffered from very human disadvantages, as well as being far less skilful soldiers. Christopher Dunphie, a National Service lieutenant with the 1st Battalion, The Rifle Brigade, had three attempts to ambush a particular terrorist; the first time he failed to see him, the second he saw him and failed to shoot him: 'I can remember him so clearly, he was wearing a blue, short-sleeved Aertex shirt, with large sweat marks under his armpits. He had a brown pork-pie hat with a wide band going round it and sweat marks coming through the hat and in his hand was a tommy-gun. It seemed to me that for an age we were staring at each other. I was completely frozen, not by any fear as to what might happen to me, but by something much more basic than that. I simply couldn't bring myself to shoot this man in cold blood.'[6]

Some months later, Dunphie had a third chance to dispose of this particular terrorist, and by that time his humanitarian feelings had given way to something else:

> I can remember, terribly clearly, shooting wildly, stupidly, and suddenly something clicking into place, and bringing my rifle out of my shoulder and becoming, in effect, a rather cold, calculating creature, putting it back in my shoulder and shooting him dead. And I think that's the moment at which I solved my own personal problem and it's also a moment which was of great value to me later on. In fact, we also shot another terrorist and when the operation was called off we went in and collected the bodies. I remember this was the first occasion that I'd actually seen a dead body and my own bullet had hit straight in the middle of the chest. It was a very small entry hole, but as I turned the body over I realized that it had hit the spine and blown the whole of the chap's back out.[7]

At which point Dunphie discovered that he had not quite become the hard man he had earlier thought, for he was promptly and most violently sick.

The MRLA guerrillas weren't the only hostiles the British Army had to deal with in Malaya – there were indigenous Malays, too – primitive people who had never left their traditional areas or given up their simple way of life – and who were not going to, simply because there was a small war going on around them. Johnny Watts, who first went to Malaya as a troop leader with 22 SAS, returned there three years later to lead D Squadron, and who later commanded the regiment, described one very lethal reminder the aboriginals left to deter would-be intruders:

> They would make a bamboo spear six to eight feet long and it would be drawn back on a whippy branch and camouflaged. To catch a pig [the trap's prime purpose], it would be set at knee height and the first thing you knew was that you had set off the tripwire and 'Shoogh!' – this thing would come out – and we had a number of chaps transfixed. A friend of mine was, through

both thighs. I mean, the force of the thing just went through one side into the other and out and he was absolutely fixed, lying on the ground. A hostile aboriginal who didn't like you [and some had been seduced into helping the MRLA] would set them at chest height and this was terrifying, and certainly slowed up your operations . . . The answer was to tie a piece of bamboo maybe a couple of feet long to the end of your rifle and make sure this was always in front of you and could set anything off. Suddenly – 'Shoogh!' and huge spears would whizz past your nose![8]

That the only effective way to combat the guerrilla fighter, particularly in the jungle, is on his own terms was a lesson the British took to heart, but one which the American military never did seem fully to learn when its turn came to fight a war under similar circumstances a decade later (though there were exceptions). Jungle fighting in the twentieth century has almost always meant close-quarter battle, with carbines, sub-machine-guns, grenades and even pistols, sometimes backed up with heavier automatic weapons and light mortars: the prime sources of firepower in a tactical situation where self-discipline and self-control combined with quick reactions and cold determination were the only rules. The Americans tried to substitute a variety of more or less inappropriate technological solutions, failed abysmally and lost a war (and over 57,000 young men) as a result; the British learned the rules quickly, and applied variants of them ruthlessly all over large swathes of what was to become the British Commonwealth as they tried to ease the transition from colonial to independent status while ensuring, wherever possible, that only those they regarded as 'good chaps' came to power. There were successes, and there were signal failures, but the 'brush-fire wars', as the professional soldiers came to call them, had one thing in common, whether they were fought in the streets of Aden or Nicosia, the mountainous desert of southern Arabia, the jungles of Borneo or the forests of Kenya's Aberdare Mountains: they were fought at close quarters, against enemies who took no prisoners.

Most of these campaigns were fought against irregular troops – not that that means they were in any way deficient in arms and equipment or fighting skills – but the British Army's second involvement in the Far East, just across the South China Sea from

Malaya, in Borneo, between 1962 and 1966, turned into limited warfare against the regular Indonesian Army. By that time, the British Army itself had also changed its character – it was now back to all-volunteer status, with the demise of National Service.

Initially, the war in Borneo seemed like a rerun of the Malayan campaign, and once again, it was the Gurkhas who provided the stiff backbone of the British forces. The main difference between it and the Malayan campaign was that an international border presented an inviolable defence for the Indonesians who ventured into Sabah and Sarawak – all they had to do was run for it, and they were safe. Until the British decided simply to ignore it.

Nominally, the cross-border operations, run under so-called 'Claret' rules, were limited to about 16,400 metres – the maximum range to which air-portable, and thus very mobile, 105mm guns could reach to give fire support to the infantry units engaged. They were limited in scope, too, with the objective of forcing Indonesian units to pull back from the immediate area of the border strip. 'Claret' operations were instigated by General Walter Walker, Director of Operations, Borneo, who hedged them around with strict conditions:

> I laid down a number of rules, which were fairly restrictive and which meant that the Commanding Officer himself had to plan the operation in very great detail. I then went down with the objective of picking as many holes in his plan as I could, because I was told quite clearly that on no account could a helicopter go across the frontier to evacuate any wounded and on no account was a dead body to be left behind, which could be photographed and shown to the world, or a wounded man taken prisoner . . . I did not allow it to be launched unless I was quite certain that it would be a hundred per cent successful.[9]

To start with, only Gurkha units were permitted to mount 'Claret' operations, but, later, British troops who had already served a minimum of one operational tour in Borneo were allowed to participate, too. Bruce Jackman's C Company, 1st Battalion, 2nd Gurkhas, was one of the first to go into action, from a base named Bakalalan in northern Sarawak; it took months to get official approval, and Walker proved hard to convince of the operation's viability, though he certainly thought the target worthwhile:

General Walker was convinced, but required more information; distances, nearby camps, what was the reaction time, all sorts of things that as an inexperienced twenty-three-year-old I hadn't really thought of and required a lot more scurrying about on the other side of the border. Then the C-i-C came down and he seemed very pro the idea and said it was clearly a good target but I had to convince everyone that it was a feasible one and that there wasn't going to be a nonsense. And so the next three to four months I spent gaining as much intelligence as we could about the place. At one stage I went and sat very close to the camp in daylight with a cine camera and I filmed sentries changing over and activities within the camp ... All the time we were training. We built replicas of the dugouts that I could see in the camp. They had no barbed wire but they had bamboo fencing around the camp so we had to devise a way of getting over it, flattening it in an assault. We practised hitting bunkers with 3.5-inch anti-tank rounds and following up straight away with soldiers getting into the bunker. So all this was going on until finally the news came that I was allowed to go. I picked the next new moon, which was ten days later.[10]

In the final hours before the assault, the plan was once again called into question after an unexpected contact with an Indonesian patrol, following which the Indonesians moved a mortar into the very place where Jackman had intended to place his mortars and heavy machine-guns. But after much internal debate, the young officer decided to go ahead:

The fence was pulled down by the first soldiers who immediately swarmed over and then the firefight started. I was enveloped in black smoke and dust from the rocket launchers and grenades, and also section commanders were throwing smoke grenades to provide some cover for themselves. There was a tremendous amount of shooting, a lot of it ours – difficult to say how much was the enemy – but there was just a huge amount of noise that will stay with me for ever. The soldiers were quite magnificent. Section commanders had got tremendously tight control within this smoke and confusion and dust and I remember seeing one, Deo Bahadur, immediately in front of me. He gripped his machine-gun group and told them to fire at a bunker while he grabbed four riflemen and ran round to one side and did a classic little section attack on the bunker, all in a very small space because the camp was about seventy-five metres across and fifty deep.

There were eight bunkers in all, four in the front and four at the back and the mortar pit, and it was quite incredible watching these soldiers diving in. Grenades were thrown in first and you saw the roofs of bunkers just lift off after the grenade explosion and there was shrapnel flying in all directions, and then kukris were out and they were in. And the speed with which it happened surprised me. I thought it would be a sort of fairly deliberate fight through. It was a fight through, but the speed that the soldiers moved from one objective to another, from one trench to another, from one tree to another, it was all over in what appeared to be seconds. Then we were through the position and the platoons had regrouped in their specific arcs.[11]

But the engagement was far from over yet, as Jackman soon discovered, because even as the Gurkhas hit them, the Indonesians were in the process of withdrawing to more secure positions on the far side of a river which ran along the back of the camp. Jackman called for cover fire from his machine-guns, sited on a hilltop and hidden by mist, but firing on fixed lines, on bearings which had previously been measured during earlier reconnaissance. 'Normally, a tracer bullet is one in five, but we had one-for-one tracer and when I called for the machine-gun fire I saw two jets of red spray almost come out of the cloud and then just go straight down the riverbed exactly where I needed it, wonderfully reassuring.'[12]

But the Indonesians now began to bring down mortar fire from their new positions, and caught Jackman and his party as they were crossing open ground:

I was faced with a sheet of flame and tremendous explosions, three simultaneously. I just felt as if I had run up against not a brick wall, but a train coming the other way, because I felt myself lifted off my feet and flung backwards. I had a terrible singing in my ears, I couldn't hear anything and all I could see was red. It cleared quickly and I found my orderly pulling my foot saying, 'Sahib, you're all right.' I was covered in mud from head to foot and what had happened was that three Indonesian 60mm mortar rounds had landed almost simultaneously, the first about seven paces in front of me and two either side. We were straddled, the four of us, and we were lucky in that we were running across a newly dug cultivated patch . . . The earth had absorbed most of the impact of the mortar rounds, which had blown upwards as opposed to outwards.

Anyway, that was a shock and it galvanized my MFC [mortar fire controller] into action, Corporal Birbahadur Gurung, who without needing any instructions darted forward to a place where he could see more clearly. He could hear a mortar firing, he spotted the source and then started to direct our mortar fire in a counter-bombardment. And it was quite extraordinary, because the first round he put down was about a hundred and fifty metres off. He made a quick correction using the knuckles on his hand and the second round actually hit the mortar pit. We saw the mortar go in the air and three bodies fly out. 9 Platoon were facing that and it was like a goal had been scored at Wembley, because a great cheer went up from the soldiers.

At this point the enemy opened up with a 12.7 anti-aircraft gun which fires a huge slug at about bomb-bomb-bomb-bomb speed as opposed to the rattle of a machine-gun [in that, Jackman may be mistaken; from his description the gun in question seems to be of a heavier calibre, perhaps 20mm]. So then we reckoned that we had about three Browning machine-guns against us plus this 12.7 from a position just over the other side of the river, about four hundred yards away, and all our concentration went now on quelling this fire. I redirected the support group with their machine-guns to a position where Lieutenant Sukhedo [commanding the support platoon] could bring his four machine-guns to bear on that site . . . At the same time the MFC started to bring down tremendously effective mortar fire.[13]

But it was having little effect on the enemy heavy machine-gun, and Jackman soon assembled a 3.5-inch Rocket Launcher team, gave them four rounds and sent them out onto the flank with orders to see to it:

About ten minutes later there was this terrific boomph from this river bank, and an explosion in among the trees, and the 12.7 stopped. It was in a bunker with overhead protection, and Corporal Lok Bahadur, the Rocket Launcher team leader, had put his first rocket straight through the slit. So it was all over. Suddenly there was a deathly hush . . .[14]

Following lessons they had learned in Malaya, the British relied to a great extent on the ability of helicopters to transcend the difficulties of operating in mountainous jungle where there simply were no tracks, let alone roads. Small units, particularly from 22 SAS, were occasionally dropped by parachute into the jungle canopy – a very dangerous activity indeed, for the trees reached a height of sixty metres and more. There were numerous fatalities as troopers shucked out of their parachute harnesses and descended to the ground on climbing tapes. More crucially, it was the SAS men, too, who bore the brunt of long-endurance cross-border patrolling, tasked with locating Indonesian positions so that 'regular' infantry like Jackman's Gurkhas, operating from fortified patrol bases just inside Sabah and Sarawak, could neutralize them.

One such base was Plaman Maku, located in the very south of Sarawak. Big enough to house a company of infantry, the fort there was quite typical – about 900 metres back from the border, so that its forward listening and observation posts stood less chance of being detected, and located on a hilltop. It had its Command Post at the centre, surrounded by sleeping accommodation, and relied for its main defence on three machine-gun-armed strong points. From February 1965, it was home to B Company, the 2nd Battalion, Parachute Regiment, functioning there as conventional infantry and exercising a hard regime of ten to twelve-day patrols followed by three or four days' defensive duties to rest and recuperate. Thus, there was likely to be less than a third of the company's 120-man strength 'at home' at any one time, and that was the case in the early hours of 27 April, when the Indonesians struck. Most of the private soldiers were eighteen- and nineteen-year-olds who had barely completed their jungle warfare training. Company Sergeant-Major John Williams was to play just the sort of pivotal role men of his rank and experience are often called upon to deliver *in extremis*, a hero's role worthy of a true Red Devil:

I was woken by gunfire in the early hours of a pitch-black morning, with the monsoon rain falling in buckets; a barrage of artillery, mortar and rocket fire was falling on the three machine-gun posts. I leapt out of my hole in the ground with my boots, slacks and belt order on, and bumped into one of the machine-gunners, who had taken two rounds to the head and who had his pistol in his hand, firing it all round him. The bullets had creased his skull and had sent him a little bit crazy and he was yelling 'They're in the position! They're in the position!' In fact, he stuck the pistol in my stomach, saying 'You're one of them!' and was going to press the trigger before I managed to take it from him. It was unbelievable because what the machine-gunner said was true. In that first barrage of fire one of our two mortar positions was destroyed, along with half the people who were manning the mortar. They had killed two soldiers and wounded several others, which brought our number down to eighteen who were on their feet and able to fight.[15]

Williams made his way to the CP, and persuaded the artillery FCO (fire control officer) to call down fire on the half of the position that the enemy had taken in their initial assault before going in search of the one platoon commander present in order to organize a counter-attack:

The position was laced with fire, it was still pouring with rain and very, very muddy. Visibility was down to about five or ten yards and the only illumination was coming from the incoming tracer rounds and the shell fire and mortar rounds coming into the position. All the soldiers facing the assault had either been killed or wounded so all the fire was incoming because there was no enemy coming at the other positions and, very professionally, the soldiers in the other sections were not wildly firing their guns. I ran across the open ground under fire to the slit trench where I knew the platoon commander to be and told him he had to bring his soldiers with him and we would mount an immediate counter-attack against the Indonesians who were in the position. As I was leading the officer and his men back across the position a mortar bomb exploded in the middle of the section, badly injuring the officer and half his section, so we were left with five men with which to counter-attack the Indonesians – there must have been about thirty of them in the position. I shouted across to one of the other section commanders, Corporal Bourne, and he was able to give me supporting fire as we put in the attack. The situation was very confused, but I shouted to the men around me that anything that came in front of them was enemy. There then followed a savage close-quarter battle with these Indonesians. One shot when one was able to but in nine cases out of ten it was hand-to-hand stuff, actually one-to-one combat. It was very frightening but it became [a matter of] survival . . . because one knew that if one didn't manage to push them out of the position or kill them, then they would kill you and overrun the position.[16]

With the Indonesian attack driven off, Williams began to try to consolidate to face the next wave. This, of course, was the position manned by the 'half-crazy' machine-gunner, and thus, the General Purpose Machine-Gun (GMPG) there was unmanned. Williams ran to it:

I remember thinking to myself, 'I've got to get more ammunition onto that gun because that's the only weapon that's going to keep the Indonesians out.' How I got there, I don't know, but I did, and I managed to lace onto the gun several belts of ammunition. At this stage, I was covered in mud from head to toe, my hands were slippery and muddy, it was pitch black, but I knew where the belts were and I knew what I had to do. One would imagine that you couldn't do it in those circumstances but it was as though you were doing it in a barrack room and you just latched them on. I then lifted the machine-gun off its pivot and directed it against the Indonesians as they came across the position. When they discovered that the machine-gun was working again they diverted the attack and about thirty of them, a platoon strength, I would imagine, assaulted the bunker. What demonstrated to me the professionalism of the men assaulting us was the person who got closest to me. I killed him when he was three yards away from me and, next morning when we were removing his body for burial, we found that he had obviously taken part in the first assault, because his left thigh had received two bullets and had been tourniqueted. It must have been very, very difficult for him to stand, never mind run up a hill, which to me epitomizes the professionalism of the TNI, the best trained of the TNKU (North Kalimantan Liberation Army) special forces.[17]

It was certainly no more than CSM Williams himself was displaying – his gun was hit by three rounds, as was a radio next to his head, and then he himself received a head wound, a serious one, which removed a part of his skull along with one of his eyes. Even so, he not only kept up his steady defensive fire, but, once the attack had petered out, and artillery fire from the neighbouring support base gave them a short respite, was able to go in search of other wounded men and gather them into the comparative shelter of the CP. 'I found young McKellar,' he recalls. 'He had his head half-severed by a piece of shrapnel and all I did was close the two halves of his head and tie it up with a bandage and tell the cook, who himself had been wounded while serving the mortar, to hold him because somebody should be with him when he died.'[18] Then yells and screams from the perimeter told him that the next attack was imminent. A quick inspection of the perimeter revealed that there were now just fourteen men able to fight, and Williams decided to try to pre-empt an assault, rather than holding one off:

I knew that should a further determined assault be made our chances of surviving it were very, very slim. We'd been fighting for about an hour and a half – although it went in a flash – and dawn was not far away. Our only hope lay in surviving to first light and waiting for the helicopters to come in and relieve the position. So we were masters of our own destiny at this time, and we had to repel any further attack. I got Corporal Bourne and his section with a couple of boxes of thirty-six grenades [those same, familiar Mills Number 36 bombs, first delivered to British troops in 1915] and we started throwing the grenades into the dead ground where I thought they were forming up, about forty yards downhill from where we were. We could hear the screams and shouts, so the grenades or the mortar [the company's mortar was being used in an almost vertical position, allowing the bombs to fall just thirty or forty yards away from the baseplate] or the [105mm] gun [from the neighbouring position] was having its effect. I think it was the gun which saved us from a third attack.[19]

Not long after, the first relief helicopters arrived, with Gurkha infantrymen and a medical team. 'We were all in various stages of dress or undress,' says Williams. 'I was still only wearing my trousers, my belt order with ammunition and water bottle and my boots and the others were similarly dressed. They looked a very sorry bunch, with blood coming from their wounds and field dressings, but the other thing they all had was that everyone had their guns in their hands. And it was at that stage that I said "We've come through," and it suddenly washed over me. And the people who'd done it were eighteen- and nineteen-year-old soldiers, not battle-hardened veterans but eighteen- and nineteen-year-old boys.'[20]

187

It took six months for CSM Williams to resume his duties, and at his reappearance on parade he was awarded the Distinguished Conduct Medal for an outstanding act of bravery and leadership. Fifteen months later, the four-year-long war with Indonesia was virtually over; almost amazingly, it had cost the lives of just 114 Commonwealth soldiers, along with 36 civilians – a death toll which should perhaps be set alongside that of the Vietnam War which was even then coming to its height, when one considers the relative merits of the two very different strategic and tactical methodologies employed.

The sort of epic defence at close quarters which the men of 2 Para mounted against the TNKU at Palman Mapu is a recurring theme in British military history, but few believed that small detachments of the British Army would continue to be called upon to sell their lives dearly in this way once the age of technology had dawned; the majority was proved wrong, once again, spectacularly and under most unusual circumstances in 1972, in the remote, broken-down coastal town of Mirbat in Dhofar, to the south of Oman. As the climber 'Brummie' Stokes, one of the SAS troopers posted there, said: 'I didn't know where the Oman was, but it was abroad. That was what mattered.'

•

Hard-earned experience of counter-insurgency operations in the 1950s and 1960s taught the British Army that guerrilla forces could only be defeated by elite troops who fought in the same way as their opponents. The guerrillas could only be defeated by undermining their support among the local population. Swift but surgical violence was also necessary to neutralize pockets of resistance, which threatened to undermine the campaign to win over the population. The Special Air Service (SAS) Regiment was such a force. SAS members were selected with great care to ensure they were up to this physically demanding form of warfare in some of the world's most demanding climatic zones: tropical jungles, mountain ranges and deserts. They also had to be psychologically prepared to operate on their own or in small groups, often far from headquarters.

The basic SAS unit is the four-man patrol, which contains experts in key counter-insurgency skills: radio communications, field medicine, local languages and improvised explosives. In the Malayan, Borneo, Aden and Oman campaigns, SAS patrols spent weeks and months living deep in guerrilla territory gathering intelligence on the enemy, winning over the local population to the British cause and then raiding enemy positions.

In Oman's mountainous and desert terrain, the SAS favoured weapons with a long reach, such as L42A1 sniper rifles, 7.62mm General Purpose machine-guns, .50 calibre Browning heavy machine-guns and M79 grenade launchers. The American M16 5.56mm assault rifle also attracted SAS users because its light weight meant extra ammunition could be carried compared to the heavy weight of the standard British Army L1A1 Self-Loading Rifle (FN FAL variant) and its 7.62mm ammunition. These weapons gave the SAS the long-range firepower necessary to pick off their more numerous Communist opponents at great distances. This was particularly important because the terrain offered little cover from observation or fire. So whoever got the first shot in usually won an engagement. Success also went to those who dominated the high ground. Here the fitness and endurance of SAS soldiers came into its own. They would often make long marches in sweltering heat to occupy key high ground by long but covert routes, giving them a dominating fire position. The surprise appearance of the SAS was often enough to make the enemy flee in disorder without firing a shot.

SAS patrols carried all the equipment, ammunition and weapons kit they needed for the duration of their patrol. Food and water were to be obtained from local sources.

Operating in small groups, the SAS set out to make contact with local Omani tribesmen, to win them over to the government cause. Rebels were persuaded to change sides and join the local pro-government militia, the *firqat*. This was dangerous work because the SAS men had to live among tribesmen whose loyalty could never quite be taken for granted. They had to be alert at all times, watching for signs of betrayal. When action did occur, the SAS patrol had to be in the thick of it to show an example to the *firqat*. Any sign of weakness by the SAS patrol could cause the *firqat* to lose heart or even change sides.

The first party of SAS soldiers to travel to Dhofar numbered just nineteen men; their task was to 'advise' the Sultan of Oman on how to defeat a Communist-inspired rebellion which threatened to depose him. In an attempt to disguise themselves, the troopers operated as the British Army Training Team (BATT), and their patrols were designated Civil Action Teams, and always included a paramedic. As well as basic reconnaissance, their job was to win the 'hearts and minds' of the Dhofaris, while the BATT's main occupation was to raise and train the *firqats* to combat the Marxist Dhofari Liberation Front – the fighters of which were known to the SAS as adoo. By mid-1972, the SAS was present in Dhofar in some strength, but at Mirbat were represented by just nine men, Captain Mike Kealy and eight troopers, with a mixed force of 55 Dhofari Gendarmes (DGs) and local *askari* levies, concentrated in the BATT house and two stone and mud forts to the north-west of the town, the whole of which was surrounded by a perimeter fence. As well as their personal weapons, the small contingent possessed a Second World War-vintage 25-pounder gun and an 81mm mortar. Just before dawn on 19 July, they were attacked without warning by between 250 and 300 adoo rebels, supported by mortars, 75mm multiple rocket launchers and heavy machine-guns. The SAS troopers were awoken by the far-away 'crump' of mortars discharging and the rush and explosion of incoming bombs. This was not entirely unusual, but none the less, they stood to, Corporal Bob Bennett* and 'Fuzz' Pussey manning the mortar and ready to lay down a protective shield of smoke. Pete Winner* was at his post on the BATT house roof with a .5-inch Browning heavy machine-gun (HMG) and next to him Geoff Taylor and Roger Coles were manning a GPMG. At the same time one of the two Fijians in the detachment, Sergeant Labalaba, dashed to the gunpit near the DG fort, 500 metres away, to bolster the single Omani on duty there, while his compatriot, Corporal Takavesi, together with the unit's medic, Tommy Tobin, began checking weapons and opening ammunition boxes. No sooner had Mike Kealy begun to take stock of the situation from the BATT house roof than it became clear that this was no routine mortaring, but the prelude to a serious assault. On hearing the first incoming mortar rounds, Pete Winner, 'Soldier I' in a later and rather more famous SAS operation:

> . . . leapt out of bed, pushed past Fuzz, Laba and Sek and scrambled up the half-pyramid of ammunition boxes which served

* Note: It is unwise to accept the names as given here as real, save for those of Kealy, Labalaba and Tobin, who are all dead, and Morrison. For example, 'Winner' has also been called 'Chapman', and 'Bennett' has also been called 'Bradshaw'. 'Takavesi' is also rendered in a number of different ways in other accounts. Real ranks, apart from Kealy's and Morrison's, are a matter of some speculation too.)

as a ladder up to the roof . . . I threw myself behind the .50-calibre Browning . . . One minute fast asleep, the next under attack, I drew a sharp breath and cursed softly, my left hand closing instinctively on the first incendiary round protruding from the ammunition box. I snapped open the top cover of the .50-calibre and positioned the ammunition belt on the feed tray. The belt held a mixture of incendiary and tracer rounds in a ratio of four to one, the incendiary rounds designed to explode on impact. With the links uppermost, I manoeuvred the belt into position with my left hand. With my right hand I closed the hinged cover and cocked the action, with a single practised twist of the wrist, feeding an incendiary round into the breech. The cold metal of the trigger felt comforting to the touch as I took up the first pressure, released the safety catch and stared in disbelief at the scene unfolding before me . . .[21]

There was no doubt now that they were under full-scale attack, the opening mortar barrage being supplemented by incoming rockets and heavy machine-gun fire. Then came the sound of the ancient 25-pounder and soon the mortar at the BATT house joined in. Winner and Taylor opened up, too, and in the illuminated darkness could see that their machine-gun fire was effective; dozens of adoo stopped, staggered and went down, some of them quite literally cut to pieces by the sustained-fire weapons at a range of just a few hundred metres.

But there were literally hundreds of them, and apart from the impressive – but actually, pitifully thin – defensive fire from the BATT house, and the steady thump of the 25-pounder, there was little to deter the adoo from either of the other two defensive positions. And with a long perimeter, much of it out of sight, it was only a matter of time before the SAS team was outflanked.

Receiving a half-garbled radio message from the gunpit, Takavesi divined that something was wrong there, and set off at a run across the 500 metres of semi-open ground towards it, desert boots round his neck and an SLR (Self-Loading Rifle; a 7.62mm semi-automatic weapon) in his hand. Almost miraculously, notwithstanding a certain evasive skill he had once shown on the rugby field, he made it unscathed, and found Labalaba trying to load and fire the gun single-handed, part of his chin shot away by a rifle bullet, the Omani gunner insensible with a bullet in the abdomen.

Sek steadied himself, the rasping in his throat slowly subsiding. Through a mist of perspiration, he surveyed the chaos in the gunpit. Laba was firing on his own. He had looked around briefly, nodded in ackowledgement, pointed to the unopened ammunition boxes and turned back to the gun . . . . The big gun continued to belch flame. It looked brave and defiant, spitting death at close range. It wasn't called the artillery machine-gun for nothing – and in the hands of Laba and Sek, the adoo must have thought it was belt-fed! They were now working as if possessed, shovelling the shells into the breech like madmen. Open breech, slam a shell in, ram it in with baton, fire. Open breech, eject hot, spent cartridge-case, kick it away. The same mechanical sequence repeated time after time after time.[22]

Minutes later – it may have been as few as five or as many as ten – Takavesi himself was hit, first in the chest, by a round which lodged perilously close to his spine, and then in the scalp; it was a superficial wound, but, as all scalp wounds do, bled profusely. In any event, his chest wound certainly prevented him from continuing to load the field gun, and he slumped to one side, picked up his SLR and commenced to fire at the enemy with that. Their position was perilous. Far too close to the by now ineffective perimeter fence for comfort, they risked being overrun as soon as the adoo realized the change in their situation. Labalaba saw one hope – a 60mm mortar which lay to one side of the gunpit. He went for it and, outside the protection of the 25-pounder's steel shield, walked into a hail of small-arms fire. A 7.62mm round from a Kalashnikov assault rifle took him in the throat, and killed him outright.

Winner, who doubled as the team's radio operator, had earlier got a 'heads up' message out to the BATT headquarters near Salalah, and now, at Kealy's instruction, sent a direct request for close air support and a helicopter to evacuate casualties. Kealy's worst fear was of penetration from the north, the sector controlled by the fort where the Dhofari Gendarmerie was located, alongside the 25-pounder gunpit, and where only the efforts of the two Fijians, it seemed, had prevented a breakthrough. Now the gun there had fallen silent. Kealy decided to go and see for himself what had happened, and detailed Tommy Tobin, the team's medic (not by any means a non-combative assignment in the SAS) to accompany him. The two followed in Takavesi's footsteps (though unlike him, they took the time to put on their boots first), while Coles slipped out of the back door of the building to place a SARBE (Search-and-Rescue Beacon) down towards the beach, to define the casevac helicopter's landing zone. (There is some confusion at this point: Winner, who was on the BATT house roof, and therefore not a first-hand witness, claims that Labalaba had died before Kealy and Tobin reached the gunpit; other reports suggest he had not, that the lull in the 25-pounder's firing pattern was simply a response to a lull in the assault, and that he was killed in Kealy's presence.)

Kealy and Tobin's run to the gunpit was a more frightening affair than Takavesi's had been, for by now grey daylight was seeping through the clouds. They reached the halfway point by the time the adoo spotted them and opened up with concerted small-arms fire. The worst danger came from a 12.7mm machine-gun, but Winner succeeded in neutralizing that threat before it material-ized, zeroing in on it as soon as it opened fire. Tobin and Kealy flung themselves over the sandbag parapet of the gunpit, the latter to land squarely in the remains of an eviscerated gendarme. Takavesi was still fighting back with his rifle, but it was clear that Kealy had arrived just in time, for already more than a dozen adoo had breached the perimeter fence and were approaching the position. Tobin went over to where Labalaba lay and began to examine him; seconds later, he was shot, too, the 7.62mm round blowing off all of one side of his jaw. (Trooper Tobin died two months later, from a secondary infection caused by one of his teeth having been driven into his chest by the original wound.) Kealy settled down to fight off the assault with rifle and hand grenades, aided by the weakening Fijian, and came very close to serious injury or death – and the one, in the circumstances, would certainly have led to the other – several times, once when a grenade landed within metres of him and failed to detonate. He was awarded a brief respite when the first pair of Strikemaster ground-attack aircraft arrived, and strafed the adoo with cannon and rockets, but they were hampered by poor visibility. Kealy, increasingly desperate, called Bennett and asked for support from the mortar, demanding that he land the rounds almost in the gunpit, so close were the attackers. Bennett could only comply when Pussey, with the mortar already up to maximum elevation, picked up its bipod legs and held it while it was fired.

Eventually, more pairs of Strikemasters arrived, and, as the visibility improved, registered greater and greater success against the adoo, finally containing them in that sector. Elsewhere, how-ever, the perimeter fence was pierced in a number of places, and the situation still hung in the balance, despite the number of casualties the rebels had suffered. The situation was exacerbated by Winner's .5-inch Browning starting to malfunction. Purely by chance, B Squadron, of which Kealy's detachment was a part, was on the point of being relieved in Dhofar by G Squadron, and the fresh troopers had arrived just days before. They had been about to set out on an orienteering exercise when Winner's first warning message, that the Mirbat team was coming under attack, was received, and immediately a troop (roughly the equivalent of an infantry platoon in size) was put in a state of readiness for deployment. When the call for close support came, they responded, too, and arrived in two helicopter lifts in time to turn the tide of battle against the Dhofaris by means of aggressive counter-attacking. G Squadron's commander, Major Alistair Morrison, wrote in his report: 'I was speechless when I saw the area of the [DG] fort. There were pools of blood from the wounded, mortar holes, many rings from grenades and the 25-pounder itself was badly holed through its shield. The ground was scarred by the many grenades which had exploded. It was obvious that an extremely fierce close-quarter battle had been fought there.' Unofficial after-action reports indicate that the rough half-circle of bodies in front of the gunpit was some five to six metres from its parapet.

The British were not alone in having an empire to lose after the Second World War stood the world on its ears: Belgium, France, Holland and Portugal all had vulnerable colonies overseas, and one by one they succumbed to pressure, most of it from the extreme left of the political spectrum, to throw off the 'yoke of colonialism' and become independent – though in some cases certainly, they were actually swapping one form of colonialism for another. Frequently, the departure of the colonial power led to a savage civil war – like those fought in the Congo and Angola. One of the earliest and most far-reaching conflicts was set in train by the expulsion of the French from Indo-China, after a desperate war which dragged on for years and culminated in the crushing defeat administered by the Viet Minh at Dien Bien Phu. Even after the departure of the French and the creation of Cambodia, Laos and the partitioned Vietnam, however, there was no real peace in that corner of the globe as pro- and anti-Communist factions continued to struggle for supremacy. Slowly but surely, the United States of America, now very self-assured in its role as the advance guard of the free world's fight against Marxism, with the might of North Korea so effectively quashed during a three-year-long undeclared war, became embroiled in South-East Asia, largely due to its political right wing's contention that the infamous 'Domino Theory' would see nation after nation fall into the Communists' bailiwick once just one had gone. The result, for America's national pride, and for hundreds of thousands of her people, was little short of catastrophic.

Chapter Six

America's Asian Wars

The United States got involved in the Korean War rather as an out-of-condition middle-aged man might rush to help a frail neighbour attacked by a teenage gang: with little thought to its own preparedness and a false sense of power and security. As a result, it got severely mauled in the first round, and only its superior weight got it through in the end. Of course, the 'police action' as it came to be known in political circles (there was never a formal declaration of war made by either side, and neither, when it was over, was there ever any peace treaty), was not an exclusively American affair – it was prosecuted, at US urging, under the auspices of the United Nations, and other countries, notably South Korea itself, but including Britain and fifteen more, also contributed forces. But America called the shots, both literally and metaphorically, and took the brunt of it, just as it was to do so in certain subsequent United Nations combat operations, for better or for worse.

On 25 June 1950, North Korean troops crossed the 38th parallel, the border between the North and South Korean states created at the end of the Second World War when Japanese troops north of that line surrendered to the Red Army, and those to the south of it capitulated to the Americans. On 1 July the first UN ground forces – American troops deployed from Japan – arrived in South Korea, but in insufficient force to halt the invasion from the north. By the end of July, all the country save for an area around Pusan, in the south-east, was in Communist hands.

General Douglas MacArthur, the American supreme commander in the Pacific during the Second World War, was given command of the UN forces, and by 1 October had driven the North Koreans back across the border, thanks to a bold plan which encompassed an (almost unopposed but none the less very risky) amphibious landing at Inchon, in the north-western part of South Korea, close to the capital, Seoul, and a break-out from the Pusan perimeter. MacArthur then ordered another landing, this time at Wonson on the eastern side of the peninsula, near the border with China, and subsequently took the UN front line almost as far as the Yalu River, Korea's border with China. The North Koreans, heavily reinforced by Chinese troops, counter-attacked, and drove the UN forces south again before the the end of year, but soon ran out of steam and were themselves driven back once more; that scenario repeated itself in April of the following year, the UN troops finally stabilizing on a line thirty to fifty kilometres north of the old border. At that point, the war settled into a static phase very reminiscent of the Western Front in 1915–17, with barbed-wire entanglements, trenches and deep dugouts marking the opposing armies' Main Lines of Resistance, as front lines had by now become known, and odd-sounding names like Imjin, Koto-Ri, Chosin and Pork Chop Hill passed into the public domain.

The advance elements of the US Army which were mobilized to turn back the Communist invasion of South Korea were almost all Regular Army soldiers from garrison units in Japan. Their weapons, equipment and tactics had changed little since 1945, so they were ill-prepared to face the Communist tanks and human-wave assaults that swept towards them.

Around half the officers and NCOs were Second World War veterans, while the enlisted troops were largely poor country boys who wanted to see the world. America's monopoly of the atomic bomb meant few of them actually expected to fight. Platoons were organized round rifle squads and machine-gun squads. The basic squad and platoon weapons were the M1 Carbine, M1 Garand Rifle, M3 sub-machine-gun, the Browning Automatic Rifle (BAR) or light machine-gun, .30-calibre Browning belt-fed machine-gun and 2.36in Bazooka anti-tank rocket launcher.

Korea's mountainous terrain meant infantry combat often took place at close quarters, with the bayonet and grenade playing a major role. Platoons tried to defend high ground where possible, to give them good fields of fire and to slow down Communist human-wave charges. Whoever held the high ground could dominate the valleys and channel enemy advances into defensive zones, and it was considerably harder to throw grenades uphill.

American platoons established close defensive perimeters, with every slit trench in sight of its neighbour. Korean labourers would usually be brought forward to dig the trenches and move supplies. Trip flares and booby traps would be put out down the slopes from the perimeter to give warning of Chinese attacks. If the enemy set off one of these devices, the platoon would man its firing positions and wait for the attack. Every man would be given an individual field of fire to cover, so there would be no gaps in the wall of lead put up by the defenders. Firepower was the Americans' main method of defence. When ammunition ran short, then US troops usually withdrew.

Liberal use of firepower also dominated American offensive tactics. Browning .30 cal and .50 cal machine-guns and mortars would saturate enemy positions for hours until the infantry would move forward. The machine-guns would keep firing on the enemy until the infantry got close enough to drop grenades into their positions. Any Chinese troops who were left alive were despatched with the bayonet.

The Allied withdrawal in April 1951 saw many hard battles fought as individual units sought to hold up the Communist advance across a broad front, but none was harder fought, nor with more honour, than that undertaken by the 800 men of the 1st Battalion, the Gloucestershire Regiment – the Glosters, as they had long been known – on the Imjin River. Over four days, the Glosters held up the Communist troops – estimated at 80,000 in one account – advancing on a main north–south axis until, their food and ammunition exhausted, virtually all the survivors, 19 officers and 505 men, marched into captivity, not to be released – save for the 34 who died in Chinese prison camps – until two years later.

The Glosters' adjutant, Captain (later, General Sir) Anthony Farrar-Hockley, proved to be not only a gallant and skilful soldier but also an author of some distinction, as his account of the battle, *The Edge of the Sword*, demonstrates. Right from the outset of the war, much of the fighting in Korea had an almost surreal quality, an illusion fostered largely by the human-wave tactics employed by the Chinese forces. Now, almost a year on, in the gathering darkness of the evening of 22 April, the Communists were still trying to batter their way forward by brute force alone, like a sledgehammer against the rapier-point of the defences of units like the Glosters:

As minutes passed, as hours grew upon the clock, I thought most often of those watching eyes along each section, each platoon, each company front; of all the eyes that peered into the darkness to our west and to our east, battalion flanked by battalion, Geordies, Belgians, Riflemen, Puerto Ricans, Americans, Turks, ROKs. And down by Gloster Crossing, where the river splashed against the fifty-gallon drums that marked the ford, sixteen men of 7 Platoon were watching especially as the moon rose over the broken walls of the village and lit the black waters.

A voice said, 'There's someone at the Crossing, sir.' Guido looked across to the north bank and saw that four figures had entered the stream and begun to cross, moving clumsily as the current caught at their knees, their thighs, their waists . . .

Now the first crossers . . . had been caught up by a further three. . . . Now their figures were more distinct, their caps and tunics moonlit, their weapons outlined. Soon they would leave the water, already receding up to their knees, and step up on to the shore and then ascend the cliff, up through the cutting.

No other sounds yet break the stillness . . . No cry of fear, no fleeing feet, no shot in panic disturbs the April midnight quiet. These men are resolute.

No cough, no careless cigarette, no clatter of a weapon carelessly handled alarms the enemy's approach. These men are trained.

The seven moonlit figures still come on; each second adds another detail to their faces, arms, accoutrements. Another twenty yards, fifteen, now ten, now five, now. . . .

Now!

The light machine-guns fire. . . . The seven men are gone, swept away, lifeless, by the fast-flowing water; except for one, poor wretch, whose groans are dying away in the shallows as the bloody water washes over him, foam-spread and ruffled lightly from his last, faint breath.

But there are more coming, hailing, many screaming as they run into the water, firing their weapons, splashing; careless of noise now that they see from whence the crossing is opposed. Seven, seventeen, twenty-seven, thirty, more than thirty stumbling and heaving their heavy limbs against the current's drag, panting and excited as they try to bear down by weight of numbers the ambuscade whose total strength theirs far exceeds. The echoing shots are now all theirs. A sub-machine-gun – homophonous, the 'burp' gun – is fired until its magazine is empty, when another sends its charge of bullets to the cliff tops, and another splits the mud and wattle of the empty village walls. There is no lack of fire from the attackers, and their comrades on the northern bank provide support; heavy and light machine-guns fire from bank to cliff across the water; mortars throw their streamlined bombs up into the night. Flash follows flash; the air above the unheeding river trembles with explosions.

Still there is no reply from the patrol on the hilltop. From somewhere to the south, their ears detect the sounds they have awaited; and, almost quicker than the thought, the shivering noise of shells is overhead as our artillery fire descends with all its might upon the northern bank and shores. In each succeeding sudden flare of the exploding shell is seen the last black veil of smoke from those preceding it. The mortars and machine-guns falter, quieten, die away. But now, again, the crossing party move in shallow water. This is the moment. Again the weapons of the ambuscade are used: light machine-gun, rifle, Sten gun, find a target each among the yelling figures underneath the cliffs; grenades are hurled among the leaders. The light of the full moon is temporarily augmented by the flares from our light mortars.

Confusion now appears among the enemy: of those that stand unwounded, some draw back; one yelling figure waves them on from deeper water near midstream; others remain in doubt, shouting in high-pitched voices from the cover of scattered boulders in the shallows, arguing back and forth. Two wounded men, whose unsure feet are turned towards the northern bank, are swept, quite suddenly, downstream, their weakened limbs incapable of fighting with the river. The ambush party's weapons find new marks among the indecisive ranks now scattered below them. New flashes and explosions rise again to force decision on the enemy, whose few remaining numbers rush in panic back across the river, the last shots of the defenders cutting the stragglers down into the water, to be swept away westward through the Imjin's mouth into the Yellow Sea.

Now there is a lull. The men in ambush examine their weapons, their commander checks the ammunition . . . A few soft phrases are exchanged; a drink of water taken; some black smudges of burnt powder removed from a cheek or forehead. The April night seems warm to these men.

Upon the other shore, the enemy is formed up again to force the crossing at all costs, the numbers of the force now ten times greater than the first small group.

200

Again, the tiny dark figures are illumined by the moon as they descend the northern bank. The word is passed along the ambush lines. Again the gunners fire with terrible effect. The mortar flares burn brightly over the black water. The lull is over; battle is joined once more.[1]

Some distance away, at battalion headquarters, the Colonel's 'O' Group was in session, reports coming in thick and fast, not just from the sub-units involved in the firefight down by the river, but from all the other components of the battalion, too, signalling their readiness for action. Then from the ambush party, a signal more important than the rest:

'They're still trying to cross in hordes, sir,' he [the battalion Intelligence Officer] says to the Colonel, who has just come in. 'In another five minutes, he [Guido, the ambush party's leader] reckons they'll be out of ammunition.'

'Tell them to start withdrawing in three minutes,' he says. 'Guy [the artillery liaison officer], I'm going to ask you for one last concentration, and then start dropping them short of Gloster Crossing as soon as the patrol [i.e., the ambush party] is back at the first cutting south of the river.'

. . . Guido's ammunition is down to less than three rounds per rifle, to less than half a magazine for the Sten machine carbines and the Brens. With the number against them, this will not hold another attack. But the task is nearly done as Guido looks at his watch. The second hand moves upwards, the minute hand closes on its next division. One after another, sixteen shadows slip south out of the village. Already some of them have formed a little covering party at the bund while their comrades go past, when they, in turn, slip softly back under cover of just such another party at the first cutting. Noiseless, as shadows are, they slip back across the unsown rice paddy calmly, as they have fought, they withdraw. Back to their slit trenches to continue the fight from there. Behind them, the body of the enemy remains unharmed – though he has a bloody nose.[2]

And that was only the overture . . .

Not unnaturally, few incidents in the action which followed were as one-sided as this; soon the Chinese found another way across the river, and attacked the Glosters' main positions from the flank in the darkness:

A faint, incomprehensible sound is heard in the night; the air is ruffled lightly; an object falls near, by a slit trench, smoking. Less than two seconds pass in which the occupants regard it, understand its nature, duck and take cover as it explodes. This is the first grenade: the first of many.

Echoing now, the hill is lit with flame that flickers from above and below. Mortars begin to sound down near the Imjin and the call is taken up by those that lie to the south behind C Company. Slowly, like a fire, the flames spread east and west around Castle Hill; and east again across the village of Choksong, as the enemy from Gloster Crossing, tardily launched at last, meets and is repulsed by D Company. Now, almost hand to hand, the Chinese and British soldiers meet. Figures leap up from the attacking force, run forward to new cover and resume their fire upon the men of the defence who, coolly enough, return their fire, as targets come to view, as attackers close with them. Occasionally an individual climax may be reached in an encounter between two men when, only a few feet apart, each waits to catch the other unawares, sees a target, fires, and leaps across to follow his advantage.

And now, to the defenders' aid, the carefully planned defensive fire is summoned. The Vickers guns cut across the cliffs and slopes by which the Chinese forces climb to the attack. Long bursts of fire – ten, twenty, thirty, forty rounds – are fired and fired again: the water in the cooling jackets warms, the ground is littered with spent cases. The mortars and gunners drop their high explosives in among the crowded ranks that press on to the hill slopes from the river crossings.

Such are the enemy's losses that now and then there is a brief respite for the defence as the attackers are withdrawn for reinforcement. The weight of defensive fire is so great that the enemy has realized that he must concentrate his strength in one main thrust up to each hilltop. As the night wanes, fresh hundreds are committed to this task, and the tired defenders, much depleted, face yet one more assault.[3]

Castle Hill, the highest point of the Glosters' forward defences, was taken after six hours of unrelenting fighting, and this gave the Chinese a decided tactical advantage, machine-guns dominating the nearby units. It was clear that a counter-attack to retake the hill was

US Marines bring mass fire power to bear during the assault on Iwo Jima in February 1945. The Americans were always prepared to use firepower rather than the bayonet to win their battles.

A flame thrower is used by US Marines to finish off a Japanese suicide squad in a bunker on Okinawa. The unwillingness of the Japanese to surrender often made such tactics the only way to deal with them.

Australian troops man a frontline position during a bitter Korean winter. Allied trench lines were often overwhelmed by Chinese human wave attacks.

In Vietnam US troops return fire against a Viet Cong sniper using overwhelming firepower. In major firefights the jungle canopy would be shredded by the hailstorm of bullets.

Left: A US Marine Gunnery Sergeant instructs recruits on the finer points of bayonet fighting. The Corps was a strong advocate of closing in on the enemy to kill him.

Below: US Army flamethrower tanks were a very popular way of clearing Viet Cong sniper positions from around American firebases because they saved casualties among assault troops.

Below left: A US soldier inspects a Viet Cong booby-trap trench. Such devices inflicted far more casualties on US troops than enemy fire.

Left: A US Marine dashes for cover during a search and destroy operation. He is wearing the lightweight body armour introduced during the Vietnam War to reduce casualties from mortar and artillery fire.

Below: British Paratroops move wounded Argentine POWs off Mount Longdon after the feature was captured in a night assault. Argentine morale collapsed under the combined weight of British firepower and aggressive tactics.

Right: A British Paratrooper stands guard with a 7.62mm General Purpose Machine Gun during the Falklands conflict. The weapon's long range and heavy rate of fire devastated numerous dug-in Argentine positions.

This Russian soldier was captured by the Afghan Mujhadeen and went over to their cause in September 1982. The Afghan superior fighting spirit proved decisive in numerous engagements with the Soviet Army.

Israeli troops cautiously move round the streets of Beirut, trying to work out where the next PLO sniper will open fire from.

PLO fighters in Beirut in 1982 used their superior knowledge of the back alleys of Beirut to inflict heavy casualties on Israeli troops who were often afraid to move away from their armoured vehicles.

The SAS assault on the Iranian Embassy in London in 1980 was a classic demonstration of the house-clearing combat skills of the British elite unit.

British troops take cover behind their armoured vehicles as an IRA bomb explodes in the Smithfield area of Belfast. The provision of specialist equipment to the security forces saved many lives during the conflict in Northern Ireland.

British troops evacuate a comrade injured during a riot. Lightly armed snatch squads were used to arrest ringleaders of mobs or protect wounded members of the security forces.

A patrol of British troops head out into the 'bandit country' of South Armagh to set up a heavily armed observation post to monitor terrorist activity.

US troops swooped into Somalia in December 1992 as peacekeepers but soon found themselves locked into bloody street battles with the forces of rival Somali warlords.

Left: US Military Police set up roof-top guard posts to protect allied positions in a Saudi city from possible sabotage by Iraqi agents or other radical elements, during the build up to the 1991 Gulf War.

Below: The ruined Somali capital, Mogadishu, was a rabbit warren of buildings that tested the urban combat skills of US troops to the full.

US Marines rehearse house-clearing tactics in the run-up to the Gulf War. The collapse of Iraqi resistance in Kuwait City meant that the Coalition forces entered the city unopposed and with minimal casualties.

US Navy SEALs board a suspected Iraqi embargo-busting ship during Operation Desert Storm. They are cautiously advancing, checking for booby traps before moving inside the ship's wheel-house.

necessary, and the officer selected to lead it, a young lieutenant named Phil Edwards, assembled men for the attempt:

> It is time; they rise from the ground and move forward up to the barbed wire that once protected the rear of John's platoon. Already two men are hit, and Papworth the Medical Corporal is attending to them. They are through the wire safely – safely! – when the machine-gun in the bunker begins to fire. Phil is badly wounded: he drops to the ground. They drag him back through the wire somehow and seek what little cover there is as it creeps across their front. The machine-gun stops, content now it has driven them back; waiting for a better target when they move into the open again . . .
>
> Phil raises himself from the ground, rests on a friendly shoulder, then climbs by a great effort to one knee.
>
> 'We must take the Castle Site,' he says; and gets up to take it.
>
> The others beg him to wait until his wounds are tended . . . 'Just wait until Papworth has seen you, sir . . .'
>
> But Phil has gone: gone to the wire, gone through the wire, gone towards the bunker. The others come out behind him, their eyes all on him. And suddenly it seems as if, for a few breathless moments, the whole of the remainder of that field of battle is still and silent, watching amazed, the lone figure that runs so painfully forward to the bunker holding the approach to the Castle Site; one tiny figure, throwing grenades, firing a pistol, set to take Castle Hill.
>
> Perhaps he will make it – in spite of his wounds, in spite of the odds – perhaps this act of supreme gallantry may, by its sheer audacity, succeed. But the machine-gun in the bunker fires directly into him: he staggers, falls, is dead instantly; the grenade he threw a second before his death explodes after it in the mouth of the bunker. The machine-gun does not fire on three of Phil's platoon who run forward to pick him up; it does not fire again through the battle: it is destroyed; the muzzle blown away, the crew dead.[4]

Lieutenant P.K.E. Edwards was awarded the Victoria Cross for his gallantry, as was the battalion's commanding officer, Lieutenant-Colonel James Carne, for his overall conduct in the battle. Carne survived the two years of imprisonment which were to follow the

Glosters' forced surrender despite being singled out for particularly brutal treatment. The battalion as a whole received the President of the United States' Unit Citation.

The Glosters' fate was inevitable, given that the odds against them, in purely human terms, were perhaps as great as a hundred to one. They were simply swamped, their courage subordinate to their supplies, their will to fight drowned by superior numbers, for the Chinese were relentless:

> We read so often in histories, sometimes in news reports, that one or another company of such-and-such a Battalion was 'cut up', 'cut to pieces', 'destroyed'. What does this mean, exactly? How is this done to a hundred-odd men, organized and armed to fight a semi-independent battle, with all the means to hand of calling down artillery fire to support them?
>
> In Korea, in the battle of the Imjin River, it was done like this.
>
> The whole Company front is engaged by fire – fire from heavy machine-guns from ranges in excess of two thousand yards, machine-guns well concealed in hollows or behind crests which take our artillery fire from other, more vital tasks if they are to be engaged. There is fire from mortars and from light machine-guns at a closer range. Meanwhile the enemy assault groups feel their way forward to the very edge of our defences; and, finding the line of our resistance, creep round our flanks to meet each other in the rear. To the defenders, these circumstances do not constitute disaster. Holding their fire for sure targets; exploiting their advantage of positions carefully sited for just such attacks as these, they are undaunted. Again and again, they see the shells from their artillery burst among the crowded ranks of the enemy. Their own small arms pile up another dreadful score of casualties. Each enemy assault is beaten off without great difficulty. For hours, this repetition of attack and repulse continues, the night wanes, the dawn begins to break. Little by little, a terrible fact becomes apparent to the men of the defence. This is not a battle in which courage, tactical and technical superiority will be the means to victory; it is a battle of attrition. Irrespective of the number of casualties they inflict, there is an unending flow of replacements for each man. Moreover, in spite of their tremendous losses, the numerical strength of the enemy is not merely constant, but increasing.
>
> Now the courage they have shown through the long hours of

the battle during the night is surely surpassed. It is no longer a question of personal bravery in the heat of the battle. This courage is now the force which makes them determined to continue an action which, for them, is almost certainly hopelessly lost. What else could they do? Why, they might run away – a number might well escape in the confusion which abounds upon a battlefield like this. They could surrender, throw their arms down, call for quarter. Surely, they might say, we have fought well; the end is the same, at all events; for us the battle is over.

But they do not. In spite of the prospect, they continue to fight with their customary skill. Being the men they are, no other course is open to them.

Thwarted, the enemy changes his tactics. Attacks continue all round; but now greater numbers concentrate on one small segment of the circle. The fire directed on to Geoff's platoon is intensified. Sergeant Robinson, who has controlled the fire of a light machine-gun group throughout the night from an exposed position, falls. His arm and shoulder shattered by a burst of fire, he none the less remains at duty, giving the crew their orders in this new engagement till he loses consciousness. Robson, a light machine-gun Number One, keeps firing though his leg is badly wounded, till he too becomes unconscious with another wound. One by one, the strength of the platoon is reduced, the little fighting groups split up. Regardless of the fire which pours down in support from the remainder of the company, the Chinese now rush in upon the few survivors – and the ground is lost. Two weak platoons and the company headquarters prepare for the next blow . . .[5]

By the morning of 24 April, after thirty hours of such fighting, B and C Companies of the 1st Battalion, The Glosters, combined, made up a strong platoon. At dawn the next day, as the remainder of the battalion was forced to surrender, what was left of D Company broke out and withdrew; they numbered just five officers and forty-one men.

Some of the US units in Korea had similar experiences in defence, particularly, as the war went on, from Chinese troops infiltrating their lines at night. Their situation was often made more difficult by the relative inexperience of their junior officers. The chief US Army historian from the European Theatre of Operations, General S.L.A. Marshall (who had also had extensive experience in the

Pacific, prior to the invasion of France) went there as an operational analyst in 1950, and returned as a correspondent for the *Detroit News* in 1953. Marshall, who made a practice of interviewing troops in groups as soon as they came out of combat, had earlier come up with many disturbing facts, the most important of which was that considerably less than 50 per cent of infantrymen ever took aimed shots at the enemy, and that many never fired their weapons in anger at all. This, to a country whose short history had been virtually built on completely erroneous, almost fabulous, tales of derring-do out of the 'Old West' during the settlement period in the latter half of the nineteenth century (when newspapers frankly printed fiction as fact, and created the myths of such as Hickock, Cody, Earp *et al.* having been bona fide heroes in the process) was quite scandalous, and has never been widely accepted.

Allied infantrymen occupying the front-line trenches at first thought they were bushes, but then they started to move closer and closer. Trumpets sounded, followed by shouting and cheering. Minutes later thousands of Red Chinese soldiers were charging towards the Allied front-lines, to be mowed down by machine-guns, mortars and rifle fire.

Chinese intervention in the Korean War changed the course of the conflict, forcing the UN forces to retreat southwards at a great rate of knots. Chinese People's Liberation Army (PLA) troops were poorly equipped but well led, highly disciplined and fanatically brave. They were all veterans of the long Chinese civil war. The PLA was equipped with weapons from a wide variety of sources: American Thompson sub-machine-guns captured from Chinese Nationalist troops, ex-Second World War Japanese Arisaka rifles, Russian SKS carbines and PPSh-41 sub-machine-guns.

Chinese battle tactics were simple but effective. They had no radios so orders to attack were issued by signal flags or bugles. American artillery and airpower could inflict heavy casualties, so the Chinese became expert at personal camouflage. Whole regiments and divisions could 'disappear' into a Korean forest or valley. The Chinese soldiers carried their own food or had to forage for it so they were not reliant on supply trucks, except for ammunition. This meant they had superb mobility over the Korean mountain ranges.

Attacks were rarely carried out at below regimental strength. Once the order had been issued for an attack, for example, whole regiments would just charge the American lines. If the defences held, hundreds, sometimes thousands, of Chinese would die. The Chinese kept repeating their attacks to wear down the defences. In the winter of 1950, the Americans did not have enough troops to defend all the front and the Chinese just swarmed around their positions, surrounding and destroying isolated American and Allied positions. Night attacks were a favourite Chinese tactic, which made it even more difficult for the Americans to react to Communist infiltration attempts. In the mountainous terrain, the lightly equipped Chinese troops could move quickly around American defences, spreading alarm and panic. It was often enough for a single Chinese patrol to appear behind an American position to force whole battalions and regiments to flee south in panic.

Chinese soldiers were strictly controlled by political commissars to ensure their fighting spirit never flagged in the face of the horrendous casualties inflicted on them by American firepower. With few supplies and in brutal winter weather, the Chinese were reduced to scavenging clothes, ammunition and food from dead or captured Americans. The Chinese soldier was expected unflinchingly to obey orders to attack, attack and attack again. He was given no option; all he had to do was point his rifle at the enemy and charge. Survival was by luck alone.

With the benefit of hindsight, it is clear that the US Army had not made the connection between the situation in Korea and that in France during the First World War, and had not, in consequence, studied the tactics of static fighting using the latter days on the Western Front as a model. When Marshall's reports from Korea began to arrive, it was clear that little or nothing had been done, either, to correct the basic problem of motivation which Marshall had detected during the Second World War, and far worse, that the current crop of junior officers often had very little idea of the mechanics of good leadership in combat, and that that had a pronounced knock-on effect. Due to the US Army policy of regularly replacing experienced troops by new recruits, to 'spread the load' of the war, so to speak, this situation had not improved even by the Korean War's last year, 1953: 'Line captains told me,' wrote Marshall in 1954, 'that morale had so far deteriorated that when units came under full attack more men died [after] taking refuge in the bunkers than from fighting their weapons in the trenches.' He goes on to describe the poor siting of bunkers and trench systems, and a basic failure to screen activity from the enemy, as endemic, even in very vulnerable locations and situations, while suggesting that on the Communist side, by 1953, in any event, that the very reverse was true.

In all, 'SLAM' Marshall's reports were a serious indictment of the US Army's day-to-day procedures; some historians argue that it was in Korea that the seed of defeat which flowered in Vietnam was sown, and there is little doubt that the roots of this pernicious growth were planted firmly in the poor training and general lack of motivation found in a conscript army forced to fight a war for which it perceived no personal purpose. This situation was reinforced by senior officers becoming divorced from the battle itself, and consequently requiring more of the men than they were able to give – a situation analogous to that which existed in France during the First World War. Nowhere is this shortcoming in training and motivation more obvious – or more dangerous – than in close-quarter battle.

It can be argued, of course, that since close-quarter battle is often little more than a mad fight for life, determination counts more than training anyway; certainly, many American soldiers who fought in Korea owed their lives to their own determination, frequently fighting for their lives with a handful of total strangers. Marshall went to the region they called Pork Chop Hill on 17 April 1953, and interviewed many of the survivors of Baker Company, 1st Battalion, 31st Regiment of Infantry, following a 20-hour firefight for a remote outpost code-named 'Dale':

Big Smith experienced that most wretched of moments when an infantryman finds himself alone in a firefight. He ran towards Pfaff's sector where he had seen the LMG firing. Little Smith in the interim had his weapon shot out of his hands by a burst of tommy-gun [*sic*] fire through the embrasure. He was already in the trench and grenading uphill when Big Smith saw him for the first time.

He said, 'My name's Smith.'

Answered Private Smith, 'So's mine . . .'

From where they stood, they could grenade both uphill and down. Flankward in either direction, it was a good twenty-five yards to a turn in the trench line [far too long to serve the twin purposes of preventing an infiltrating enemy from enfilading and reducing the blast potential from grenades; even a cursory study of First World War history would have revealed that shortcoming]. Little Smith had brought to the spot a full case of grenades and an ammo can filled with loose grenades. He had already thrown nineteen grenades against the Chinese on the upper slope. So with his work cut out for him [*sic*] Big Smith conformed. Surrounded, they stood back to back, grenading all round the clock. Their last stand was maintained in this manner for approximately ten minutes [curiously, Marshall – who had certainly pushed out the boundaries of veracity in individual after-action reporting – often accepted individuals' conceptions of the passage of time without question. Under the circumstances, it is frankly impossible to believe that this stage of the action can have taken so long; see below]. Big Smith still had his carbine and Little Smith his M1. But because of the lack of clear sightings and defined targets, there was no pay-off in these weapons. During this stage, Private Serpa came back to them wriggling along the parapet. But he didn't join [in the] fight. His M1 had jammed and a grenade explosion had numbed his right arm.

They got down to their last two grenades. They had thrown about forty and had drawn a little blood as they knew from the flashing and the screaming. At least a dozen enemy bombs had exploded within a ten-yard circle of where they stood, but fast footwork had kept them unhurt.

Until Big Smith yelled, 'Hold the grenade!' they had fought without a word. He signalled his intentions by fitting carbine to shoulder. They had cleared the immediate slope and the surviving Chinese had crowded down into the main trench. Then for perhaps another five minutes, they held the enemy at bay, Little Smith firing one way down the trench, Big Smith spraying carbine fire in the opposite direction. In the same second, they ran out of bullets.... From Little Smith's end of the trench, a squad of Chinese charged towards them as the rifles went silent. They had already pulled the grenade pins.

Big Smith yelled, 'Throw!' Both grenades landed and exploded right amid the enemy pack. Several Chinese went down, and the others recoiled.

Little Smith asked, 'What do we do now?'

Big Smith said, 'Over the side! Give me your foot!'

He threw the smaller man over the parapet. With a vault, Big Smith landed right behind him. Together they rolled about fifteen yards downhill. There, Big Smith grabbed his partner by the leg and held him. Serpa had rolled with them, and was already flattened.

'What do we do now?' asked Little Smith.

Said Big Smith, 'You're dead. No matter what happens, don't move. It's the only chance.'

The new game was improvised on the spur of the moment. The three of them were to play it for the next four and one half hours on a spotlighted field rocked by shellfire from the ridges held by friend and enemy.[6]

Even as close as forty-five metres away, the outcome of the Smiths' firefight – even its existence – had gone completely unmarked, for the good and simple reason that communication and organization had broken down completely within the platoon (and also within the company and battalion of which it formed a part); the whole unit was split into small groups of men fighting for their lives with no time even to spare a thought for regrouping and trying to fight as a unit again. Normally, such a breakdown in order is fatal.

Dale's defense was already disintegrated, though Pfaff, Reasor and a few of the other diehards didn't know it. The [Third] platoon became lost in the few minutes when the Communists had passed the big trench and mounted the knob on scaling ladders. It mattered very little that, combined with the American VT [variable time – a type of artillery shell fuse designed to detonate a shell just above ground level] fire, their own artillery came in on the Chinese, killing many of them in the open before they could strike a blow with their hand-carried weapons. The sufficient number remained quick and moving, to abet the shellfire in its work of destroying the defense. From the enemy platoon which had won the high ground, one squad peeled off and, running downslope, entered the main trench between Pfaff's squad and the platoon CP.

Of this entry Pfaff was unaware as he ran in that direction trying to pry his comrades from the bunkers so they would fight for the trench. Sergeant Reasor and his scratch squad had already been driven back from the CP area by an enemy wave which swept up Moore's Finger [an outlier to the ridge on which Dale stood] and into the trench, thereby completing their envelopment. But even Reasor's situation remained unknown to Pfc. Frank Minor and Pvt. Randolph Mott who together were defending a fighting bunker just ten yards from the CP and not thirty-five yards down the trench from Reasor's group. Both men were having their first bath of fire.

So swiftly had come the on-fall that it seemed unreal to Mott. He wasn't dazed, he felt no fear but he saw no reason to get excited. When his rifle [an M1 Garand] jammed from a ruptured cartridge after he had fired a few rounds down Moore's Finger, he dropped the weapon and fell in beside Minor to serve as BAR [Browning Automatic Rifle] leader. Minor was shaky but doing his best. Mott figured a little close assistance would steady him. Neither man knew the Chinese already held the height above the bunker since no one came to tell them.

Mott was shocked out of his lethargy by a heavy explosion from the direction of the CP. Mott's own bunker shook from the impact [*sic*]. After the action, platoon survivors said, 'Two artillery shells hit the CP roof together.' But there was no way of knowing this; the blast that hit the CP could have come from a hand-thrown satchel charge. Next Mott heard a man screaming in pain. It roused him as nothing else had done. He knew his leader hardly at all. But something told him that the cry came from Bressler and that the lieutenant was mortally wounded.

He went from the bunker door and looked down the trench. The CP was in ruins, its roof caved in and its nearest wall half-down. A Chinese stood at the broken entrance shining a flashlight inwards. Mott saw at least five other Chinese with him. Oddly then, instead of grabbing the BAR, he yelled to Minor, 'Hand me a cleaning rod! The Chinese holding the flashlight heard the shout and turned the flashlight on Mott. He ducked within the bunker and with one violent thrust cleared his weapon. As he made the door again, he saw the enemy group still bunched outside the CP. He emptied the weapon and saw three men fall. Two others ran directly past him before he could pull again. The sixth Chinese jumped for cover behind the sandbag superstructure of Mott's own bunker. Like most other works of the kind along this grotesquely engineered American front, the bunker loomed so high above the outwork that an enemy gaining the wall was in a perfect position to enfilade the trench.

This boarder had no such aim. Mott nipped into the trench trying to catch his silhouette against the skyline. But the man was already over the roof and flattened. Leaning down the face of the bunker, he tossed two grenades in the embrasure. They exploded under Minor and shattered his legs. An engineer sergeant *who until then had been sitting quietly in a corner* [italics added] bestirred himself to help Mott lift Minor into a bunk.

Reaching for the BAR, Mott found it dry of ammunition. So doing, he stumbled across a pile of filled sandbags. He yelled to the engineer, 'Help me fill the window!' While they heaved at the bags and stuffed the embrasure, three more grenades came through. Mott fielded them with a quick pitch through the bunker door. As he clutched the third one, touch told him it was American issue. He scooped it up and outward in one motion. It exploded outside the door frame in the faces of a second Chinese party rushing the bunker.

That inspired him to search the bunker for more grenades. He found a full case. Then he stood at the door grenading toward the mouth of the chogi [communications] trench which led back to the main line. Targets were plentiful. By that time the Chinese in large numbers had climbed the saddle to the rear of Dale Outpost and the back door was bending to their weight. While his grenade supply lasted, Mott would fight. When it ended he would plug the bunker door from the inside, as he had done the embrasure, and try to stay the night.

Private Pfaff's sortie down the trench line, made to rally the other men in the same minutes when Mott started functioning in the fight, proved futile. In the first bunker there were five of the platoon's Koreans, each armed with a BAR. None was firing. Their inertia was robbing the defense of a great part of its most useful firepower.

Pfaff yelled, 'Come out and fight!' Not a man stirred. He tried again. 'Come out or I'll kill you!' They pulled back to the far wall and were in the darkness. Two ideas blocked his impulse [to shoot them]. He wasn't sure he had authority, and if he spent bullets there, he'd disarm himself. So he walked on down the trench. Within a few minutes the Koreans were grenaded to death in the bunker where Pfaff had left them paralysed by their own fear.

At the next bunker, Pfaff saw a Chinese setting up a heavy machine-gun on the roof. He fired once with his carbine and the man pitched backward down the slope. A second Chinese took his place and a second bullet got him.

Passing the CP bunker and seeing that the roof was down, Pfaff sang out, 'Anybody in there?' No answer came back.

Possibly his voice was drowned by the battle sound. Pfaff simply took it that 2nd Lieutenant Ryan Bressler and his assistants had moved elsewhere in the outpost. At the fifth bunker he joined Pfc. Dwight Marlowe who sat nonchalantly on the roof firing a machine-gun down Angel Finger [another spur] as if he enjoyed the work. For two minutes they sat back to back. There was no conversation. Uphill, 30 feet away, three Chinese armed with burp [sub-machine] guns were trying to dust Marlowe from the roof. The fire was going over his head. Pfaff fired his carbine just long enough to silence them. It distracted Marlowe not at all. He kept the LMG firing downslope and turned only long enough to turn to Pfaff and yell, 'Go to the next bunker; they need help!'

At the door of the next bunker stood Pfc. Rivera Rodriquez firing down the trench toward the chogi trail with his M1. Within the bunker were five engineer privates who had come to the hill in early evening to work on fortifications. As the fight started, a satchel charge put through the embrasure had mangled all five men; three were unconscious. Rodriquez, with a bullet through one arm, was trying to protect them. Pfaff dressed Rodriquez's wound crudely and then said, 'Cover them a little longer. I'll get more dressings at the CP.'

He could barely force his way into the ruin. The thin air was heavy with dust and powder fumes. A thin voice called to him, 'Help GI, help GI . . .' He knew it was Bressler. He crawled along the floor. A second voice, closer to him, whispered, 'Help . . . help.' He could barely hear it but he crawled toward it. In the darkness he fell over the body of Sgt. Chapman Spencer. He was legless from the hips down.

Bressler called to him again. 'Come to me!' Weak, the voice was still insistent. He left Spencer and crawled toward Bressler. From the doorway a Chinese opened fire with a tommy [sic] gun. The bullets splashed among the fallen timbers as he crawled. He got to Bressler. The lieutenant's legs were pinned by two of the fallen beams. Pfaff strained and sweated to free him. The effort was useless. Sandbags from the ruined roof pinned down the beams.

He tugged on. Three hand grenades came through the door and exploded in arm's reach of him. The same beams which pinned Bressler insulated the explosion, and saved his life.

Bressler said, 'Quit it now; you can't help me.'

Pfaff said, 'Then I'll stay with you.'

Bressler said, 'You are to go [and get help; specifically to request more artillery support into the outpost position itself]. It's an order.'

There was a small exit on the far side. Pfaff crawled towards it. As he made the opening, a wounded Chinese, moving on all fours along the outside wall, tried to raise his tommy gun. Pfaff was carrying an unpinned grenade. He rolled it, and the explosion blew the man apart.

Once more in the open, Pfaff was intent only on carrying out the order, and in his agitation clean forgot his promise to return to Rodriquez. He ran for the chogi trench. Hardly had he entered it when a squad of Chinese blocked his path. He threw a grenade and without waiting for the explosion, jumped from the trench and went down the slope in a diving roll. By sheer luck, he landed in a shallow gully which ran under the concertina [barbed-wire] entanglement. Then he started for the company position on the big hill. His feet were like lead and he could hardly summon the energy to move them.

By now the fight had been going just a little less than one hour. Mott had walled-in his bunker. The two Smiths and Serpa were playing dead under the lights. [Save for Marlowe, alone at his machine-gun] only Reasor and his men still stood their ground. But they were no longer firing. For about thirty minutes, they had maintained steady fire around the circle with M1 and carbine [the M2; basically, a modified M1 capable of automatic fire. It was a widely unpopular weapon in Korea, by virtue of its innate unreliability]. The concentration kept the Chinese either under cover or at a respectful distance. But at the end of the half-hour, all weapons were out of ammunition.

Their predicament was eased only because by then the Chinese in the vicinity were in the same plight. Reasor saw them wandering back and forth across the knob obviously searching for something to fire. They were in such numbers that they could have crushed his group by sheer physical weight had they grouped and rushed the trench. But they moved like sheep, seeming too dazed even to seek defilade [cover] against the artillery fire. It was agonizing to the Americans to see these targets moving freely within less than twenty yards of where they stood, helpless to cut them down . . .

The nine men were wholly surrounded. The trench was blocked in both directions. There was nothing to fire and no one paid them the slightest heed. Flares brightened the ground all round. Shell-fire continued to rock the hill. They simply sucked in their guts and hugged the shadow of the trench wall. For at least two hours they stayed that way, silent and motionless.

Pfaff, by now exhausted, took two hours to reach the company CP, arriving there just after midnight. The company commander had already called up a small party under Lieutenant George Hermman to reinforce the Third Platoon at Dale, and now he increased its size and sent it on its way. Fresher than Pfaff had been, after his experience during the firefight [and with a longer downhill, and shorter uphill, stretch to negotiate] they made it to the chogi trench approaching the outpost in an hour. Almost immediately they found themselves under attack from three sides: by Chinese reinforcements climbing towards the trench on both sides, and by a third which blocked the entrance to Dale itself.

Private John Dawson yelled, 'The Chinks are on our right!' and Pvt. Esequiel Leos answered, 'They're on our left!' Corporal Shuman, opening fire with his BAR, aimed straight along the trench, dissolving the enemy group in the foreground. With rifles and grenades, the men to his rear opened fire down the slope. The Chinese weren't routed, but after a few men died the others sought protection behind the rocks. Four of Hermman's men were hit by the countergrenading. For the moment the passage became quiet and the platoon surged forward.

Hermman got up to Mott's bunker. Mott heard him yell, cleared the sandbags away from the door and joined the party. Shuman, Dawson, Leos and two others had meantime started down the trench on the other side of the outpost, still thinking that after mopping up a few Chinese, they would become solid with the Third Platoon.

Of the group continuing with Hermman, thirteen were new Korean replacements. Mortar shells were now ranging in on the trench line, coming in four-round salvoes. Hermman told the Koreans to continue down the trench and clean out the bunkers. They shook their heads and refused to move.

By sheer coincidence, Sergeant Reasor was deciding in these moments that if his group was ever to get out, he would have to make a break for it. He had not heard either Hermman's yelling or Shuman's movement, and he was still unaware that a support party had arrived on the hill. Reasor said, 'Follow me!' and they went. Corporal Droney had found a last clip of ammunition within the bunker and, being the only armed man, brought up the rear.

Reasor's party stumbled into Hermman hard by the CP bunker . . . Hermman, arguing with his Koreans, was yelling, 'Get the hell on! Some of his GIs were kicking at the ROKs in a vain attempt to get them started. They simply looked blank. Reasor heard a noise right behind him. Glancing upslope, he saw two Chinese setting up an HMG not more than twenty feet away. So he grabbed a grenade from a Korean and threw; it exploded at head level, killing both men.

As the sound died, four mortar shells came in dead on the group. Three exploded along the embankments. One landed in the trench. A shard pierced Hermman's leg and went up to his intestines. Three of Reasor's men were also wounded. Another round, exploding through the embankment, had sprayed fragments among the recalcitrant Koreans, wounding four and killing two.

There followed several minutes of confusion. By radio, Hermman got word that Shuman and his men on the other side of the outpost had been checked by tommy gun fire and were starting back toward the chogi trench. He called Lieutenant Jack Patteson [the company commander], and said, 'I'm hit and I've lost ten men out of twenty-four.' [Which was not, strictly speaking, correct.] He didn't say he had to withdraw but Patteson drew that conclusion, telling him, 'Fall back on Check Point One,' which was an outguard position between Dale and the main line.

So Hermman called out to his men, 'Fall back on Jackson!' that being the code name for Check Point No 1. No one had ever told Reasor about that code name. But one soldier in his group was a Private Jackson and this character was already running for the chogi trench. Misunderstanding Hermman, Reasor took out after Pvt. Jackson and two of his men followed. Split from Hermman, they collided with a Chinese party at the mouth of the chogi trench and recoiled, once more taking refuge in a bunker on the other side of Dale.[7]

At this point, Marshall maintained, the Americans could still probably have saved Dale by reinforcing it further, arguing that the artillery fire was serving its purpose and keeping large Chinese assault parties at bay, though he does accept that the artillery fire having been called off while Hermman's party made its way forward had permitted the Chinese to bring up reinforcements of their own. The withdrawal continued, Pvt. Mott leading the way, but the party came under grenade attack again, and again went to ground. One man, Platoon Sergeant Burton Ham, managed to reach the company command post on his own and then went out again to guide what was now frankly just a rescue party. Eventually, all Hermman's bedraggled and much-reduced party got back to the company position, leaving Dale to the Chinese and Reasor's small party and the two Smiths and Serpa, all of whom were either supposed dead or had simply been forgotten. Outpost Dale was soon retaken by an attack in company strength which cost further American and Korean lives, a necessity which could have been avoided if communications between Hermman and the Baker Company CP had been clearer. At the end of the engagement, 123 Chinese bodies were removed from the position. Reasor's men joined up with the attacking force; the two Smiths and Serpa – now injured again by artillery fire – were eventually rescued after lying doggo in the open for more than six hours.

Just as during the First World War, life in the line in Korea was basically one of life-threatening boredom interspersed with very occasional intense periods of close-quarter battle, and also like the First World War, one of the main distractions was to be found in patrolling in the no man's land between the opposing forces, both for intelligence-gathering purposes (though they were actually better served by Koreans who passed among those peasants who were still, almost incredibly, trying to live out their lives in the war zone), and simply to cause damage and kill the enemy. 'What's the object (of patrolling)?' one high-ranking rear-echelon officer (who still should have known better) asked S.L.A. Marshall. He replied, 'To prove you're top dog. To demoralize the other fellow as much as possible and to make his patrols feel weak in the stomach before they even start.' Sometimes, though, as we shall see, patrols found themselves in a much more critical, even pivotal position, where a cool head introduced a new and unlikely element into what was still essentially close-quarter battle – namely, the massed guns of the artillery.

Some members of the multinational force in Korea were better at patrolling than others, and, for Marshall, the best of all were the Ethiopians, who contributed but a single battalion to the force, and rotated it every year. Marshall was full of praise in general for the Ethiopians, who never contributed a senior officer to the UN Force.

217

They couldn't read maps, he said, but never missed a trail; were 95 per cent illiterate, but took over US comms. equipment and operated it twice as well as the best trained Americans, and they lost not one single prisoner, nor left a dead man on the battlefield. The Ethiopians were at home in the night, left no tracks, seemingly shed no blood and, to the Chinese, left every impression of being nothing more than deadly phantoms.

Less than a month after the loss and retaking of Outpost Dale, a new Ethiopian detachment, the Kagnew Battalion (named after the charger of King Menelik during the first war with Italy, in 1896), came into the theatre and took up a position on a ridge line, 300 metres above the valley floor, on the right flank of the US 7th Infantry Division. Before them, some 800 metres out into what was effectively no man's land, lay two outposts, 'Yoke' and 'Uncle', each – just like 'Dale' – entrenched and embunkered all around its summit, with its upper slopes liberally covered with barbed-wire entanglements. The ridge was occupied by the Ethiopian First Company, under Captain Behanu Tariau; Yoke was garrisoned by 2nd Lieutenant Bezabib Ayela with fifty-six men, the nearer Uncle by a smaller body:

As with a hundred other such small positions forward of the Eighth Army's main line, the twofold object in garrisoning Yoke was to parry any attack before the Chinese could reach the big trench, and, also, to lure the enemy into the open where he could be blasted by the markedly superior American artillery. It was wearing duty, for it made troops feel like bait in a trap. So the garrisons were rotated every five days.

Twenty-two-year-old 2nd Lieutenant Zeneke Asfaw's mission of the night of 19 May was to descend, with his fifteen-man patrol, into the main valley about 800 yards to the right front of Yoke and in this disputed ground attempt to ambush a Chinese patrol and return with prisoners. This was more or less the routine objective in all patrolling. He started his march at eleven o'clock with his second-in-command, twenty-one-year-old Corporal Arage Affere, leading the column. In exactly thirty-five minutes they reached the bottom. The actual trail distance had been one and one half miles, partway uphill where the route traversed two ridge fingers. But they had done most of it at a running walk.

At twenty minutes before midnight, having seen nothing of the enemy, Asfaw decided to halt. The patrol had come to a concrete-walled irrigation ditch. Where Asfaw stood, the ditch did a 90-degree turn, with the elbow pointing directly at T-Bone hill. To the youngster came the flash inspiration that here was the tailor-made deadfall. Three trails crossed within a few yards of the bend in the ditch. He could deploy his men within the protecting walls and await the enemy. He distributed his men evenly around the angle with one BAR on each flank.

It was done in the nick of time. There was the briefest wait. At ten minutes before midnight, Asfaw, straining to catch any movement in the darkness, saw standing in the clear, 300 yards to his front, a lone Chinese. While he looked, approximately a platoon built up on the motionless scout and simply stood there, as if awaiting a signal. It was a tempting target; though too distant for his automatic weapons to have more than a scattering effect upon the enemy force, it was still vulnerable to the American artillery fire which could be massed at his call.

Asfaw switched on his radio to call Battalion. By some fluke, in those few minutes as the game opened, it wouldn't cut through. He spat in disgust at a technical failure which, seen in retrospect, was clearly a blessing in disguise. The whole pattern of this strange fight developed out of the accidental circumstance that during the next half-hour, the Chinese felt free to extend their maneuver and Asfaw, being without radio contact, had to keep telling himself that he had been sent forth to capture prisoners.

In that time, the body confronting him swelled to two platoons, but still did not move. That meant that close to a hundred men would be confronting his fifteen. That was fair enough. So he crawled along the ditch, cautioning his men to maintain silence and hold their fire until he gave the word.

When at last the Chinese moved toward him, it was not in columns but in a V-shape like a flight of wild geese, with the point marching directly toward the apex of the [bend in the] ditch. All this time, Asfaw had been concentrating his attention on the enemy directly to his front. Now as he turned his gaze to the files at the far ends of the V, he caught what all along his eyes had missed: five hundred yards to his left, another Chinese company, marching in single file, had passed his flanks and was advancing directly on Outpost Yoke. He looked to his right. Another body of the same size had outflanked him and was marching against the ridge seating [his own battalion's] First Company. With that, he saw the problem as a whole. He was in the middle of a Chinese battalion launched in a general attack. Its grand deployment was in the shape of an M, and the V-shaped body advancing on his ditch was simply a sweep which tied together the two assault columns.

By now, they were within 200 yards of him and the column on his left was almost at the foot of Yoke. His radio was still out. To his immediate rear was an earth mound perhaps ten feet high. Thinking that the mound might cause interdiction, he moved leftward along the ditch, whispering to his men to stand steady and testing his radio every few feet.

The eight Chinese forming the point of the V were within ten yards of the ditch when Asfaw yelled, '*Tekuse!*' [fire]. He had already arranged it so that the fire from his two flanks would cross, so that both sides of the V would be taken in enfilade. The eight-man point was cut down as by a scythe. The two wings which followed at a distance of fifteen yards lost another dozen men to the BARs before the surprised Chinese could [react] and go flat. At that moment, Asfaw's radio sparked, and he raised a friendly voice. This was his message as entered in the journal: 'The enemy came. I stopped them. Now they surround me. I want artillery on White Right [the area to his left rear].' In his own hour of emergency, Asfaw was ignoring the force to his front, hoping that he would still be in time to shatter the column moving against Yoke.

No sooner had he given the direction than he saw the flattened company in his foreground start skirmishers around his left flank. He felt this was the beginning of an envelopment. Still he did not amend his fire request. Instead, he shifted more of his men leftward in the ditch, figuring that with grazing fire he could slow the movement. As he said, 'By then I had steadied, and was enjoying it.'

In three minutes, the American barrage fell right where Asfaw had wanted it. Illuminating shells from the 155s began to flood-light the valley. By their glare, Asfaw could see the killing rounds biting into the column, killing some Chinese, scattering others. But he could also see figures in silhouette moving against Yoke's skyline and he guessed that the enemy had penetrated the works. He got on the radio to report.

So he was just a bit too late for a perfect score. The small-arms fire all around him had meant he couldn't know that the Chinese had massed fire against Yoke and against the big ridge almost coincidentally with the opening of his own engagement, cutting all comm. wires and forcing the men back into their bunkers.

(The men) on Yoke ... heard the first volley of Asfaw's skirmish. But it sounded far away. The impression it made was swiftly erased when the enemy artillery deluged their own hill. Both radios were hit, and the field phone went dead. 2nd Lt. Ayela moved from post to post crying *'Berta!'* [standby], but for all the noise, he had no forewarning of what was coming.

Realization came when three red flares cut the night above Yoke's rear slope. Ayela ran that way along the trench, knowing they had been hand-fired by the enemy. At the rear parapet, he could hear voices chattering from downslope. Yoke's rear was lighted by a searchlight beamed from the battalion ridge. Raising himself to the embankment, Ayela could see at least a squad of Chinese working up through the rocks not more than thirty yards away.

Corporal Ayelow Shivishe was with him, and lived to tell about it. Within call of Ayela were thirteen riflemen and one machine-gunner, covering the backslope, but all in the wrong place to see the approach. Before Ayela could either fire or cry out, he was drawn back the way he had come by the sounds of shooting and a piercing scream right behind him. Two squads of Chinese had come up the side of Yoke, killed a BAR man and jumped into the main trench.

Ayela ran for them, rifle in hand. In full stride he was blown up by a bomb – an ordinary hand grenade with TNT shaped around it, used by the Communists to clean out bunkers. Shivishe went flat in the trench and emptied his M1 into the enemy group. He saw three men go down. Then he knew he had been wrong to waste time that way. The hill was leaderless, though no one else knew it. Shivishe ran the other way round the trench to tell Sergeant-Major Awilachen Moulte that he was in command. As he made the turn, the Chinese at the rear slope came over the parapet and were in the trench. But not unopposed. The machine-gunner, Pvt. Kassa Misgina, had heard the noise and rushed into the breach. He cut down the first three men. Then two things happened right together: his gun jammed and an enemy grenade got him in both legs. They were deep wounds, but didn't jar his fighting rhythm. Misgina passed the gun back to a rifleman and told him to get it freed. Then he grabbed a box of grenades and, returning to the step where Ayela had been those few seconds, resumed the fight to block the rear portal. Reasoning that if he kept them ducking for cover they couldn't rush, he stood on the parapet and let fly.

None of this ordeal was known to Asfaw down in the valley or to the higher commands tucked away among the high ridges. But unlike every other actor in the drama, Asfaw alone could see all the parts of the big picture. From his place in the ditch, he had witnessed the enemy's grand deployment. Also, he knew that the observers on the high ground had no such advantage, and due to the interposition of the lower ridges, could catch only fragmentary glimpses of the developing action. When his eyes told him that the artillery dropped on Yoke's forward slope had effectively shattered the reserves of the Chinese column there, his reason replied that the hills masked that fact to everyone else.

Having already concluded that his task was to destroy the Chinese battalion by use of artillery, and realizing that he only was in position to regulate fires so that there would be no 'overkilling', Asfaw saw clearly that a halfway success along either flank must finally doom his patrol. Driven back, the survivors would converge within the draw which was his escape route to the rear. There they would reinforce the company which was moving around his left flank. From that side, the ditch provided no protection.

For fifteen minutes he had watched the artillery stonk the Chinese on the lower slope of Yoke while doing nothing about the enemy column attacking First Company's ridge. The reason was that the latter force had made slower progress and was still toiling toward the hill. During the same interval the Chinese platoons to his front had continued to crawl around his left flank and were now even with his position. He took another look at the enemy's solid column on his own right rear; it was just fifty yards short of the main incline. At that point he gave his direction, calling for fire to be placed where it would catch the attack head-on. It was delivered 'on the nose'. The column attacking First Company began to dissolve and recoil toward him.

In this manner, while his withdrawal route was still open, he made his decision to fight it out on the original line. Here was a youth having his first experience under fire. But the role he had voluntarily accepted made requisite a sense of timing rarely found in a division commander. The Chinese nearest to him continued to extend their outflanking maneuver; he but shifted a few more riflemen to the left to slow them with grazing fire.

On Yoke, Sergeant-Major Moulte's first act after taking command was to run to Ayela's body to make certain he was dead. Two enlisted men had been felled by the same bomb which killed the lieutenant. Moulte yelled for stretcher-bearers. Then, gathering six men and passing each an armful of grenades, he swung along the trench toward the front of Yoke on the heels of the Chinese group which had entered the works after killing Ayela. It was a sneak movement, the men moving silently and in a crouch, with one scout five yards to the front. Surprise was complete. Perhaps thirty-five or forty yards beyond, the scout gave an arm signal, hand out, pointing the direction for the grenade shower. The explosions came dead center amid the enemy group. Some Chinese were killed. Others scrambled for the parapet or tried to hide next the sandbag superstructure of the bunkers. Moulte saw at least six Chinese in clear silhouette as they climbed up from the trench wall. He was carrying a BAR. But didn't fire a single shot. He said later, 'I don't know why. I just didn't think of it. Ordering others to carry on and hunt down the invaders in detail, Moulte doubled back to see how things were going on the rear slope. By then, the wounded Misgina's machine-gun had become freed, and with that weapon he was holding the portal, supported by one BAR man. Moulte counted ten dead Chinese in front of Misgina's gun. Beyond them, he could count at least thirty of the enemy among the rocks. They became revealed momentarily as they grenaded upward. But the distance was too great and the bombs exploded among their own dead.

From his hilltop on OP 29, First Company's commander, Captain Tariau, could eyewitness the skirmishing on Yoke's rear slope. He had received the relayed message that the patrol had engaged; it came to him during a period when First Company could no longer raise Battalion [due to another technical fault]. Then, for fifteen minutes – the critical period when Moulte was rallying his men to repel boarders – his own radio cut out. His anxiety mounted because of his helplessness. By the searchlight's glare, he could see the Chinese massing against Yoke's back door but he was in touch with no one. When quite suddenly his radio cut in again, he told 2nd Lieutenant William DeWitt, his artillery forward observer, to hit Yoke directly with VT fire and illuminating shell: 'Fire Flash Yoke Three.' [A Fire Flash order told all guns to put maximum ordnance down on the designated target.] Minutes later, Yoke was under a fierce rain of hot steel.

The effect of the fire was to drive Moulte's men back to their bunkers for protection while transfixing the Chinese in the open. There was thirty minutes of this. Then Tariau asked for a curtain barrage on both sides of Yoke to box in the enemy survivors. He pondered extending the barrage across the forward slope, then rejected the idea, apprehensive that Asfaw's men might be falling back on Yoke.

On that score, he might have saved himself worry. Asfaw was still sitting steady in the ditch and enjoying it. By now the Chinese who had been to his front had completed the half-circle and were spread across his rear. Their skirmish line, a lean hundred yards from him, had already been joined by the first stragglers retreating from Yoke, Uncle and First Company's hill. His big moment was at hand when, having nailed his colours to the mast, he would now win or lose it all. He gave the fire order which called in the artillery, 'Fire Blue Right'.

If his guess was right and the fire was accurate, this would crush the Chinese to his rear and fall just short of his own position. There were two minutes of suspense. Then the barrage dropped, dead on target, braying the enemy line from end to end. He kept the artillery on Blue Right for ten minutes; when it lifted, there was no more fire from his foreground or immediate rear.

But it had been, and remained, the closest thing. At the moment when Asfaw asked for Blue Right, his own patrol was wholly out of ammunition save for the cartridges in the magazines of three M1s. The fragments of the two main enemy columns continued to drift back toward him. He knew that the patrol's survival from that point on would pivot on the radio [which, we may recall, had already proved fault-prone] and the accuracy of his call to the artillery.

At last, fortune rode wholly with him. The fight continued on these terms for two more hours, with no firing from the patrol. There were times when the Chinese, rebounding from the two outposts and then regrouping, got within fifty yards of the ditch. Blue Right never failed him. There were also times when he asked for and got barrage fire on all four sides of his patrol, thereby closing the enemy escape routes leading to T-Bone Hill.

By four o'clock in the morning, the battlefield was at last quiet, and Asfaw could see no sign of a live enemy. The patrol rose and stretched, satisfied that it had done a good night's work. Asfaw radioed the message, 'Enemy destroyed. My men are still unhurt.

We have spent our last bullet.' Being now unarmed, the patrol expected a recall.

What came back proved with finality that Ethiopians prefer to fight the hard way. This was the message from Captain Aleu [the Battalion G2] to Asfaw: 'Since you have won and are unhurt and the enemy is finished, you are given the further mission of screening the battlefield, examining bodies for documents and seeking to capture any enemy wounded.'

That task, which entailed another four to five miles of marching, preoccupied the patrol for the next two hours. The light was already full and the bird chorus in full song when I [i.e., Marshall] met them as they re-entered the main line. Asfaw went briefly into the statistics of the fight. On the ground within 150 yards of the ditch he had counted seventy-three Chinese dead. On the slopes of Yoke and within the trenches were thirty-seven more enemy bodies. There were other bodies in the paddies forward of Uncle, still not counted. But assuming the usual battle ratio of four men wounded for each one killed [a dubious yardstick in that particular war, perhaps; see below], the score said that he had effectively eliminated one Chinese battalion.

As a feat of arms by a small body of men, it was matchless. No other entry in the book of war more clearly attests that miracles are made when a leader whose coolness of head is balanced by his daring. Victory came not because of the artillery, but because Asfaw believed in it.[8]

A decade and a half later, the United States of America found itself committed to another war against the forces of Communism in Asia, this time in Vietnam, finally abandoned by the French a year after the Korean cease-fire. The Vietnam War had little in common with the Korean – there was never a true Main Line of Resistance, as such, and casualty figures there were much lower as a result – in Korea, in three years of war, the UN forces suffered 118,515 killed in action and 264,591 wounded, plus 92,987 captured, the majority of whom never returned, having died of mistreatment and starvation (of 10,218 Americans who became prisoners of war, only 3,746 eventually returned home); in addition, an estimated 3 million South Korean civilians lost their lives. The authoritative *Brassey's Battles* puts Communist battle casualties at 1.6 million. In Vietnam, between 1 January 1961 and 27 January 1973, American forces lost 45,941 men killed and suffered another 300,635 wounded (figures for those killed and injured before the first US combat unit arrived in Vietnam on 8 March 1965 reflect casualties among 'advisers' and airmen). Viet Cong casualties during the Tet Offensive alone (during the early part of 1968) exceeded American losses for the whole war. Estimates put the total death toll of North Vietnamese Army – NVA – and Viet Cong at around 100,000, 50,000 less than the number of ARVN (South Vietnamese government forces) killed. Thus, if one goes by casualty figures mapped over time, the Korean War was a much more devastating affair than the Vietnamese, but there was one striking difference, at least for the USA: in Vietnam, the Communists won.

The Special Forces, or 'Green Berets', were among the first US troops to see action in South-East Asia. Officially only advisers, they were soon leading South Vietnamese troops into action as the Communist offensive gathered steam.

Formed in the 1950s to raise guerrilla groups behind enemy lines in the event of the Third World War, the Green Berets were in the forefront of stopping Communist revolution in the early 1960s. Special Forces units were split into so-called 'C' Detachments that worked with South Vietnamese corps headquarters. Out in the field were 'B' Detachments which commanded a variable number of 'A' Teams. An 'A' Team was a twelve-man self-contained unit that was able to operate for long periods away from base. During the early years of the American involvement in South-East Asia, the 'A' Teams were sent into the jungle interior to help set up border monitoring camps with the help of indigenous tribesmen.

These camps were spartan affairs, often little more than a few mud huts, a couple of badly built bunkers, a small airstrip and a rudimentary barbed-wire perimeter. Special Forces soldiers were mostly long-service Regular Army junior officers and NCOs. They had to be proficient in a wide range of military skills, weapon handling, first aid, intelligence gathering, languages, radio communications, unarmed combat and calling in air and artillery support. This latter skill was particularly important given the exposed position of most of the Green Beret bases. The Green Berets were some of the first units to use the M16 5.56mm assault rifle in combat. This automatic weapon was significantly lighter than its predecessor, the M14, but it suffered from teething troubles, including a high stoppage rate. This was a serious drawback in close-quarter combat.

Patrolling with local tribesmen was the main job of the 'A'

Teams, who would send out daily sweeps into the jungle to locate Communist supply trails running into South Vietnam down the famous Ho Chi Minh trail. Jungle patrolling was not a simple skill to master. Navigating in conditions where you can only see a few metres ahead was carried out by compass. Local scouts would be sent ahead and out to flanks to give warning of enemy attacks. If the scouts were good, the patrol would have time to set up an ambush. The main body of the patrol would form the killer group, covering the main ambush site, or killing zone, with all their automatic weapons. On either end of the killer group would be cut-off groups, whose job was to kill any enemy that escaped alive from the main ambush site.

If the 'A' Team and its local allies were ambushed, then their main job would be to get out of the killing area as fast as possible. In many cases Viet Cong ambushes were so well planned and executed that the only way to escape the ambush was to charge the enemy ambush group. Brave and suicidal, but it often threw the Viet Cong off their guard and forced them to pull back.

There have been endless discussions of the validity of the tactics the Americans (and their allies, notably the South Vietnamese themselves, of course, but also Australians and South Koreans in considerable numbers) used in Vietnam, and the most damning condemnation of them has to be that despite the expenditure of billions upon billions of dollars, they did not secure a victory. The ill-contained fury of senior figures in the American military establishment at this failure – which occasioned such outbursts as the threat 'to bomb Vietnam back into the Stone Age', and locally, at least, the attempt to do just that – certainly did little to help, and neither, at the other end of the scale, did a very obvious disaffection in the ranks of the 'grunts', the infantrymen from both the Army and the Marine Corps who had to bear the brunt of the fighting.

The reluctance of succeeding American political administrations to give the generals the forces they wanted was cited by most pro-war 'hawks' as a major contributory factor, but there is every reason to believe that such escalation could only have had much more dire consequences, and in any event, was against the will of the majority of Americans. The proof of the matter is that a sufficient number of Vietnamese themselves wished to live in a state run along Communist lines, and that many of them were prepared to fight – and die, if necessary – to achieve that.

The conflict in Vietnam did allow the American military to

experiment with new tactics. To make maximum use of the helicopter's mobility, for example, the US Army formed a special airmobile division in the early 1960s. In 1965 the 1st Cavalry Division (Airmobile) was dispatched to Vietnam to spearhead the growing American effort to defeat the Vietnamese Communists.

The '1st Cav' flew into action in Bell UH-1 'Hueys' and giant Boeing CH-47 Chinooks. Close air support came from Huey gunships equipped with rocket pods. Traditional artillery was flown forward by the Chinooks to bring the enemy within range. The division's infantrymen had to be specially trained to operate from helicopters, which involved air navigation, fast roping from hovering helicopters and travelling light. The division was perhaps the only one to fight in Vietnam that had time to train in peacetime before being committed to combat. It was made up of a mix of Regular Army officers and NCOs, Reserve officers serving their active duty time and draftee soldiers. Their successors would not have the luxury of peacetime training to mould them into a lean, mean fighting unit.

Newly developed personal weapons and equipment meant '1st Cav' troopers carried an impressive array of firepower. Each platoon had two of the new M60 machine-guns, which could fire 200 rounds a minute out to 800 metres. The M16 assault rifle with its automatic capability was to prove invaluable in close-quarter jungle combat. Each squad and platoon also had small, lightweight radios to call in artillery fire, air strikes and medical evacuation helicopters.

Once dropped off at jungle landing zones (LZs), it was the job of the airmobile infantrymen to seek out and destroy the elusive Viet Cong and NVA. Platoon patrols were sent into the jungle to engage and fix the enemy in combat. Once engaged, the enemy would then be destroyed by superior firepower. Platoons would generally advance in arrowhead formation in open country or single columns in thick jungles. Where possible the M60 teams would set up in fire positions to cover the three rifle squads as they advanced. If the platoon came under fire, the M60s would open fire and suppress the enemy while the rifle squads took cover. The enemy position could then be assaulted. If the platoon came under concerted enemy attack, it would withdraw into a narrow perimeter so individual soldiers could not be picked off one by one. The platoon commander also needed to know where all his soldiers were so he could confidently call down fire support from artillery, gunships or aircraft. In such situations, platoons in contact with the enemy would receive priority for fire support, and it was common for lieutenants or sergeants to adjust fire support on to targets.

This was a great responsibility because the slightest mistake could result in so-called 'friendly fire' landing on their own troops. The '1st Cav's' decentralized use of fire support meant every infantry squad potentially had the whole division's massive artillery and air assets at its disposal. In many encounters with the Vietnamese Communists this capability would prove decisive.

American offensives in Vietnam usually took the form of 'search-and-destroy' missions which relied on the mobility of helicopter-borne forces, and sometimes these turned into full-blown battles such as was the case at A Shau; but compared with Korea, casualties were light – the battle for Hamburger Hill (more accurately, Hill 875), which is rated as one of the heaviest engagements of the early period of the war, cost just 281 American lives (and an estimated 1,400 Viet Cong) despite lasting for twenty-three days.

Despite the aggressive role the Americans saw themselves playing, it was actually the Vietnamese forces which initiated most combat. Statistics quoted in a memorandum written for the US Secretary of Defense in 1967 suggested that 49 per cent of all combat incidents in Vietnam resulted from VC/NVA forces ambushing American troops, while 30 per cent were originated by VC/NVA attacks on static American positions; just 7 per cent of all firefights were chance engagements, and only in 14 per cent of all cases did US forces initiate the engagement. The ratio probably changed later, but not by too much. (Quoted in *The Pentagon Papers*.)

The Army of the Republic of Vietnam (ARVN) was the mainstay of the anti-Communist war effort throughout the conflict in South-East Asia. Drawn from former colonial units when South Vietnam gained its independence from France in the mid-1950s, it was never really a match for its Communist opponents. The ARVN was never a cohesive Western-style army, even after the US began its massive Vietnamization programme in 1969. This programme put huge quantities of US hardware in the hands of the South Vietnamese and allowed its infantry to trade in its old M14s and Browning Automatic Rifles (BARs) for first-class M16 rifles and M60 machine-guns.

ARVN battalions were usually commanded by former French colonial soldiers, who had a fanatical hatred of the Communists. Junior officers had grown up during the 1960s, when American influence was at its height, so they were Westernized to a degree. The ordinary soldiers were peasants conscripted into the army, and on occasion showed little enthusiasm for aggressive actions. Army pay was minimal and there was little official help for soldiers' families, so each ARVN base was also home to soldiers' dependants. South Vietnam was far from being a politically and socially cohesive state, so these divisions were reflected in the ARVN. Unless family or regional ties existed, it was common for officers to treat their soldiers as expendable cannon-fodder and only be concerned about their own personal welfare.

Some ARVN units, under the prompting of their US advisers, tried to use American tactics, but many had their own distinctive style of warfare. ARVN units lived off the land or carried their own livestock with them. They tried to be self-sufficient as much as possible because they could not rely on their corrupt logistic support system.

In defence the ARVN could be very determined and dogged. When defending fortified bases, for example, they had no equal, and used covering fields of fire with machine-guns and adjusting artillery or air support on to targets.

On offensive operations or patrols, ARVN troops were very cautious. Their calculating commanders, from platoon to division level, would only commit their troops when they were certain of an easy victory.

ARVN troops would rarely go on patrol in less than company-sized groups to avoid Viet Cong ambushes. Heavy casualties would impoverish many of the soldiers' families, so ARVN troops could not be easily persuaded to embark on risky operations.

In their home areas the ARVN would fight doggedly to protect their families and homes from Communist attack and assassination squads. ARVN commanders often became very experienced at predicting and countering Viet Cong activity. This allowed them to pre-empt and intercept VC units moving to attack their positions.

Almost half of all contacts with the enemy which resulted in close-quarter fighting came in the form of ambush, and the ambush was, as ever, a terrifying experience:

No one took any chances in the bush. Every time something jumped, they fired it up. The Vietnamese had what was called an L-shaped ambush. They waited until they got the main body of troops in the killing zone. I don't care how many people – there used to be about 120 of us – everybody in the killing zone was supposed to be dead in approximately seven to ten seconds, if the ambush was effective. That was why it was very important to have that point man stay alert, for the simple reason that he was the first man in the woods. The people behind him couldn't see what was happening because the bush was so thick; people in back of the point were concentrating on the flanks and watching for booby traps. But the point man was the first man through the woods, so he might have picked up a little noise or some other clue to let him know that there might be an ambush up ahead. A lotta times we used to have scout dogs, and the dogs would alert the point, cause they could smell out the Viet Cong and the North Vietnamese. Then the point . . . would simply raise his hands with his palms up to the man right behind him. Everyone would halt and the word would get back to the company commander. Everyone stopped, and each person would start alternating on the flanks; one man would face to the right, and one man would face to the left, automatically.

The ambushes were set up to annihilate you, just dust you in a hurry. The trick in an ambush was to attack the main body. If you didn't know where the last man was, then you didn't know how many people you were up against. Say you hit the point man, you still didn't know how long the column was. You might have been hitting a battalion. And everybody in that column was going to come just right on over your ass on the ambush site. But if you saw the last man come through, then you knew you could get a full count. It was proven that some units hit too soon. Sometimes you had to wait a couple of extra seconds just to make sure that that was the last man because they might have had two units running. That last man might have been maybe twenty-five feet from the beginning of the new column. That was the nerve-racking part of the ambush. You were about ten feet off the trail; everything was camouflaged, even your face. You had a bush over your head as maybe you were lying flat. You could actually count people as they came by. Once the last man was through, that was when the shorter part of the L-shaped ambush opened up; that was the signal for everyone to pop up. Usually the claymores [plastic-bodied anti-personnel mines, much the same shape as a hip-flask. Filled with an explosive charge and thousands of 7mm ball bearings, they are typically mounted waist-high, and can be command-detonated or activated by a trip-wire; they have a lethal radius of up to twenty metres. The Viet Cong had similar mines, which they manufactured locally from scrap metal and the explosive charges recovered from undetonated bombs and shells, of which there was no shortage] then frags and then the smoke – rifle platoons started raking the area. The claymore mines disorganized them because the mines were so effective. Much more effective than taking a shot with your weapon, 'cause it was going to clear out the whole area for sixty feet in front of you. The claymores were set up along the long part of the 'L' and you might even have [them] doubled up. When the claymores went, that was the first indication that everybody should have been operating.

When you were caught in an L-shaped ambush, the main thing was not to panic. I know it's easy to say 'Don't panic', but out in the bush, like I say, if you hesitated, that was it. There's two kinds of fighters – the quick and the dead. When you got in that kind of situation, there was only one thing to do. You had to charge that ambush. You had to go right over them to get out of the killing zone. After you walked into the L, that was the only way out. The L was set up so they weren't on either side of the trail – else they'd be shooting at each other. So they planted pungi sticks [short stakes, often of bamboo, driven into the ground and subsequently sharpened; they were sometimes smeared with excrement to act as an expedient poison] on each side of the trail. You couldn't fall down on the trail or else you'd be falling right into death. And if you froze and didn't know what to do, that gave them more time to fire you up in the killing zone. You couldn't go back – 120 men couldn't turn around – so you had to go right at the direction the fire was coming from. They were faced then with 120 men coming at them. But they'd been following you; whatever it took, they had it. If you were a company, maybe they were a battalion. If they were going to take on a company, they were going to have personnel to deal with that. If they got a battalion-sized force, it might have been strung out eight blocks long to meet your company-sized two-block [long] force. Right in back of the ambush, they would build bunkers. If they had to pull back, they withdrew behind the bunkers, and then the bunkers opened up to stop you from overrunning them because you blew their ambush.

Each side set up ambushes . . . Usually, if you got attacked it was because you got caught up in an ambush or you ran up on them by accident. In the area around Bao Loc ... we was [*sic*] making contact almost every night. I was scared of every little lightning bug or anything that moved. I thought it was a VC. In the day I was all right. At night, anything could happen. The CO would designate which guys would take their turn on night ambush, which could get kind of hairy. The purpose of the night ambush was so we wouldn't get ambushed ourselves. If, for instance, we had made contact that day and we had so many kills, then we knew Charlie had got his ass kicked. Could have been his brother or some friend who got killed that day. So we had to go back down the trail knowing he was following us. We went with maybe eight or ten people and were thinking about getting hit. When we went on ambush, we would go back down the trail about a half a mile. Then we'd get off the trail about fifteen feet, set up claymore mines, and in some cases we'd set trip flare wires across the trail. When they hit the trip flare wires, right then and there they'd be in our killing zone. We'd just pop the claymores which would spray an area about sixty feet straight out in front and about forty feet in width to the sides. That could be hairy, too. If the first part of the enemy was coming up the trail and they hit the trip wire and we hit them, they might have been only the point of a whole regiment. It was doubly hard to tell at night – it was pitch black. If that happened, we were in trouble. They were going to come and get our ass.[9]

After ambushes, assaults on static locations were the next most utilized tactic. The most significant Vietnamese attack on an American installation was probably the 'siege' of Khe Sanh, a large US Marine Corps firebase in the very north of the country, close to the de-militarized zone (DMZ) which marked the border with the North and also the border with Laos. Here, however, there was comparatively little close-quarter combat (though the term is definitely relative); the battle was chiefly fought in the form of artillery duels, and would-be infiltrators were dealt with by intense machine-gun fire. US Marines patrolled outside their own perimeter, but rarely ventured further than 500 metres; they had strict orders to withdraw, and not to fight, if they made contact, and call in supporting fire. There were a number of occasions, however, when NVA 'sappers' succeeded in penetrating the US perimeter, either that of the main compound or of one of the outposts, like that on a prominent ridge-line north-east of Hill 861, attacked during the early hours of 5 February 1968. At around 03.05 hours, NVA artillery began an intense barrage, under cover of which sappers and assault troops blew lanes through the perimeter wire on the north side of the outpost and assaulted the position held by Echo Company, 2nd Battalion, 26th Marines. And it was 2nd Lieutenant Donald Shanley's 1st Platoon which took the brunt of it:

> Quickly the word filtered back to the company CP that the enemy was inside the wire and Captain Breeding ordered that all units employ tear gas in defense, but the North Vietnamese were obviously 'hopped up' on some type of narcotic and the searing fumes had very little effect. Following the initial assault there was a brief lull in the fighting. The NVA soldier apparently felt that having secured the northernmost trenchline, they owned the whole objective and stopped to sift through the Marine positions for souvenirs. Magazines and paperbacks were the most popular. Meanwhile, the temporary reversal only served to enrage the Marines. Following a shower of grenades, Lieutenant Shanley and his men charged back into their original position and swarmed all over the surprised enemy troops.
>
> The counter-attack quickly deteriorated into a mêlée that resembled a bloody waterfront bar room brawl – a style of fighting not completely alien to most Marines. Because the darkness and ground fog drastically reduced visibility, hand-to-hand combat was a necessity. Using their knives, bayonets, rifle butts and fists, the men of the 1st Platoon ripped in to the hapless North Vietnamese with a vengeance. Captain Breeding, a veteran of the Korean conflict who had worked his way up through the ranks, admitted that at first he was concerned over how his younger, inexperienced Marines would react in their first fight. As it turned out, they were magnificent. The Captain saw one of his men come face to face with a North Vietnamese in the inky darkness; the young American all but decapitated his adversary with a crushing roundhouse right to the face, then leaped on the flattened soldier and finished the job with a knife. Another man was jumped from behind by a North Vietnamese who grabbed him around the neck and was just about to slit his throat when one of the Marine's buddies jabbed the muzzle of his M16 between the two combatants. With his selector on automatic, he fired off a full magazine; the burst tore huge chunks from the back of the embattled

Marine's flak jacket but it also cut the North Vietnamese in half. Since the fighting was at such close quarters, both sides used hand grenades at extremely short range. The Marines had the advantage because of their armored vests and they would throw a grenade, then turn away from the blast, hunch up and absorb the fragments in their flak jackets and the backs of their legs. On several occasions, Captain Breeding's men used this technique and 'blew away' enemy soldiers at less than ten metres.

During the fighting, Captain Breeding fed fire team-sized [i.e., four- or five-man] elements from the 2nd and 3rd Platoons into the fray from both flanks of the penetration. The newcomers appeared to be afraid that they might miss all the action and tore into the enemy as if they were making up for lost time. Even though Breeding was no newcomer to blood and gore, he was awed by the ferocity of the attack. 'It was like watching a World War Two movie,' he said. 'Charlie [from Victor Charlie, the NATO phonetic alphabet rendition of the abbreviation VC, for Viet Cong. In fact, of course, Khe Sanh was under attack from North Vietnamese regulars, not indigenous South Vietnamese guerrillas, but the Americans used the diminutive "Charlie" ubiquitously] didn't know how to cope with it. We walked all over them.'[10]

The body count, so beloved of the Americans, giving as it did an easy-to-assimilate (though frequently false) idea of the dimensions of the battle, revealed that 7 Americans had died (all of them, Breeding estimated, in the initial barrage) and 35 were wounded, while the NVA 'suffered 109 known dead; many still remained in the 1st Platoon area where they had been shot, slashed or bludgeoned to death. As near as Captain Breeding could tell, he did not lose a single man during the fierce hand-to-hand struggle ... All in all, it had been a bad day for the Communists.'[11] It will perhaps come as no surprise, thanks to its characteristically 'gung-ho' tone, to learn that the preceding account comes from the History and Museums Division of the US Marine Corps itself (and is derived from an interview with Captain Earl Breeding). It is thus certainly Corps propaganda, though that does not mean that it is necessarily inaccurate, the apparent vast disparity in the body count notwithstanding.

But what of the enemy? Just as the Marines were among the cream of the American military, so the NVA had its own elite. Sappers were the NVA's elite assault troops thanks to their superior training and aggressive fighting spirit. Sapper in Western military terminology is a military engineer who specializes in building bridges and the like. The North Vietnamese Communists used their sappers as combat engineers. It was their job to cut a path through enemy defences and then destroy key installations with demolition charges.

US bases in South Vietnam were protected by formidable defences that included claymore anti-personnel mines to shower attackers with deadly shrapnel, trip flares to light up the night and give warning of an attack, barbed-wire entanglements, deep bunkers and watch-towers. To defeat these defences required special men. The NVA treated its sappers as precious resources and husbanded them for very important operations, such as the attack on the US Embassy in Saigon at the start of the 1968 Tet Offensive.

When released for an operation by the NVA high command, the sappers went about their business in a thorough and methodical way, and then pressed home their attack with ruthless efficiency. First, the sappers would spend days, sometimes weeks, carrying out a detailed reconnaissance of the enemy defences, looking for trip wires, claymores and identifying key positions, such as command bunkers, within the American base. This was done with total stealth to avoid alerting US intelligence to the presence of sappers in the neighbourhood.

Once the plan was ready the sappers would move forward at night to take out the defences. Often dressed only in black shorts and covered in truck engine grease to make them merge with the darkness, the sappers methodically clipped the command wires on all claymores and trip flares. Lanes were then cut through the American barbed wire. With no prior warning, the American sentries had no time to react when the sappers rushed forward from the jungle and dropped demolition charges in their slit trenches. This was usually the signal for NVA regular infantry or local Viet Cong support troops to open fire with RPK machine-guns and AK-47 assault rifles on positions around the US perimeter. Mortar and sometimes CS gas shells fired into the US base added to the defenders' confusion.

With the sappers inside the American base, they would plant demolition charges in the command bunker, mortar pits and artillery batteries. As confusion reigned among the defenders, the sappers were often able to inflict heavy casualties before they withdrew back into the jungle. Once they were outside the wire, the sappers just disappeared into the jungle to prepare for their next attack.

There was another aspect to the Vietnam War which harked back to an earlier age. Like the British and Germans on the Western Front during the Great War, the Vietnamese dug tunnels – but this time, they were for use as clandestine bases, not to plant huge explosive charges. To clear them out, the US Army had no real alternative, so extensive were some of the tunnel systems, but to send men down into the darkness armed with grenades, pistols and sawn-off shotguns, and hope that these tunnel rats, as the volunteers soon became known, would be able to overpower the defenders in the subterranean darkness . . .

All in all, the war in the tunnels was perhaps the most bizarre aspect of the entire Vietnam conflict. Operating in the tunnels themselves was clearly a mind-bending business. As well as the obvious fears and dangers – Viet Cong fighters lying in wait in the darkness; the confines of tunnels built for the slighter Vietnamese frame (and often completely devoid of any form of shoring) and the disorientation of operating in what was essentially a three-dimensional maze; booby traps; poisonous insects and reptiles, and rats (some of them, it was believed, carriers of bubonic plague) – there was often another form of obstacle entirely: Viet Cong dead. One of the pioneer tunnel rats, Harold Roper, who served with the 25th Infantry Division in 1966, said, of that, and of the whole experience:

> The Viet Cong would take their dead after a battle and put them down in the tunnels; they didn't want us to count their dead because they knew we were very big on body counts. Finding them wasn't pleasant, but we'd killed them, so it didn't matter. It was worse if they'd been down there for a week – it stank! Everything rotted very quickly because of the humidity. I came across rotting bodies several times.[12]

And as for the living . . .

> . . . they used to put boxes of scorpions with a trip wire, and that was a booby trap. . . . I got bitten by a centipede once. That sucker was probably a good eight inches long and I thought I was going to die . . . and one hole that seemed to be darker than any hole I'd ever been in . . . I thought I was losing my equilibrium

because it seemed like the hole was moving in on me, and as I shined the light around some more, I found out it was just a mass of spiders . . . we used to call them one-step, two-step or three-step snakes . . . Bamboo Vipers. They weren't very long but they had a potent bite; once bitten, you could only take one or two steps. The Vietnamese somehow tied the viper into a piece of bamboo, and as the tunnel rat goes through he knocks it, and the snake comes out and bites you in the neck or the face.[13]

None the less, the worst enemy the tunnel rats had to face was two-legged:

. . . The tunnel straightened out then went another ten yards and stopped at a wall. A little dirt fell from the ceiling at the end, betraying the existence of a rectangular trapdoor to another level up. Flowers held his lamp steadily on the door.

The NVA soldier was evidently lying just over the trapdoor. Batman [Sergeant Robert Batten, a three-term Vietnam veteran] moved up beside Flowers [Jack Flowers was a lieutenant, and recently appointed as the team's leader] and made to push upwards on it. But Flowers prevented him. Flowers was Rat Six, and the point man; he insisted upon dealing with the situation by himself. Batman crawled back a few yards. Flowers tensed in apprehension; sweat was running into his eyes. He edged up to the wall and sat under the trapdoor about twelve inches above his head. He placed a lamp between his legs, shining upwards. Then he put his hand under the door and exerted a small amount of pressure. Batman cocked his pistol; Flowers gripped his. Flowers took a deep breath of the dank air and pushed up on the door. It yielded. He twisted it and set it down crosswise on the bevelled frame. Then he paused, planning to slide it away and start firing into the void.

A foot above Flowers' glistening and grimy face, the trapdoor was quietly turned around and slotted back into its frame. Flowers froze; the gook was right there. Suddenly, the door moved again. Something dropped into Flowers' lap, right in front of his eyes. He watched it fall, momentarily transfixed; then the danger to his life overwhelmed him as he screamed 'Grenade!'

Flowers recalled his thoughts from his hospital bed. 'If it had been John Wayne,' he said, 'he would have picked up the grenade, lifted up the trapdoor and thrown it back at the bastards.

If it had been Audie Murphy, he would have thrown his body over the grenade to save Batman's life ... But since it was Jack Flowers, I started crawling like hell!' In fact, both men received the Bronze Star.[14]

Though hospitalized with grenade fragments in both legs, Flowers was soon up and about, and resumed command of the Diehard Tunnel Rats, as the team was informally known. He was soon back in action:

Morton, going point, reached a trapdoor leading upward and went through it. Then he let out a piercing scream ... After the scream, the next thing Flowers heard was three shots, followed by Batman calling him. When Flowers reached the trapdoor, Batman was standing in it. Morton was rolling on the tunnel floor, his hands covering his face. Blood was flowing freely through his fingers. 'The son-of-a-bitch knifed him,' said Batman. 'I think he got it in the eye. Batman was shooting as Flowers began working Morton back down the tunnel. Morton moaned as they hauled him along, a few feet at a time. He was unconscious when they got him out. His face was covered with coagulated blood and dirt. There was a great gash starting at his hairline and running across the bridge of his nose, done through his left cheek. Flowers couldn't tell if he still had his right eye when the dust-off helicopter ferried him away.[15]

Once the 'rats' had explored a tunnel system, and determined, as best they could, its extent, they would place demolition charges, both to cave it in and to kill any Viet Cong or NVA they had been unable to winkle out, but even that was a dangerous business, and unreliable, too – a system which was believed to be destroyed might, in fact, have simply been an isolated outlier to a much bigger complex, which would now be more difficult to discover. Sometimes, though, the engineer laying the charges got it just right, with unexpected results:

They had to haul the C4, crate by crate [there were twenty-five of them, making 300 pounds in all], through the long tunnel complex and set them scientifically at the same time. Each case was a foot and a half square, and they were soon physically exhausted. Then they discovered that they had not brought enough fuse wire, only a foot and a half, which would be extremely dangerous to use on that amount of explosive. Ellis asked him, 'Swofford, is that enough fuse?' and Swofford, who had seen too many Westerns, lit the fuse with a cigarette, and answered, 'No, you'd better start running now, Elly!' One could not, of course, run down a tunnel, but professionals did perfect an astonishingly fast if undignified crawl, which stripped the skin from elbows and knees. As the two men shot out of the shaft like a couple of corks, the 300 pounds blew. Swofford was knocked down by the blast. The tunnels exploded as if in slow motion. Hundreds of tons of earth and stone hung suspended in the air. It was an unusual luxury to use 300 pounds of explosive, but it was the tunnel where Virgil Franklin had been killed . . .

After the earth had settled, they discovered Swofford had set the charges so perfectly that the entire lid of the tunnel maze had been lifted like a cap from a skull. They saw tunnels they had never even found during the early searches. Ellis forgave Swofford the theatrical gesture. Swofford loved every minute of it.

Harold Roper summed up the experience of a tunnel rat, and one may imagine that he was speaking for many, when he said: 'It didn't revolt me. I was just an animal – we were all animals, we were dogs, we were snakes, we were dirt. We weren't human beings – human beings don't do the things we did. I was a killer rat with poisoned teeth. I was trained to kill and I killed. Looking back, it's unreal. Unnatural. It almost seems like someone else did it. It wasn't really me, because I wouldn't even think of doing anything close to that again.'[16]

Eventually, the burgeoning American network of paid informers/defectors led to the discovery of more and more tunnel systems (while the Phoenix programme of clandestine assassination succeeded in targeting Viet Cong cadre members pointed out by those same informers), which in turn forced the Vietnamese Communists to rely more and more on regular NVA assets. But it was the halt to the wholesale bombing of the North, particularly of Hanoi and Haiphong, which actually sounded their death-knell, for it released the massive B-52 bombers based in Thailand and on Guam, each one capable of carrying 22.5 tonnes of bombs, to fly missions against the tunnel complexes at high altitude. Dropped in sticks 800 metres long, the hundred bombs each aircraft carried left a swathe of destruction composed of craters up to twelve metres deep, and caved in tunnels reaching even far below that. The tunnel rats – or almost all of them, anyway, for there were those who had come to love their strange subterranean way of life – breathed a collective sigh of relief, and tried to return to something approaching normality.

Chapter Seven

Through Tribal War to Techno-War

Just as the period immediately after the First World War had produced a rash of new weapons developed as a result of lessons learned in the four and a half years of war, so the 1950s saw widespread additions to the world's armouries in the light of developments between 1939 and 1945; by the 1960s, the infantry-man's equipment had changed considerably, as a result. One particular expansion was in the use – and usefulness – of grenades and mines, both command-detonated and self-tripping, a trend which was set to continue as the century progressed. Grenades, in particular, took on a new significance as the technology for project-ing them improved, and by the 1960s, the distinction between them and small-calibre mortars had blurred almost to the point of the two being indistinguishable; but true hand grenades, too, had become much more efficient weapons.

There are two basic types of anti-personnel grenade – those which rely on blast alone, and those which generate significant amounts of shrapnel. The German *stielhandgranate* (or 'potato masher', as Allied troops called it, from its distinctive shape), used through both world wars, was an example of the former, while its British equivalent, the venerable Mills No 36 bomb, was a first-generation fragmentation grenade. The two types are sometimes known as offensive and defensive grenades, respectively, since their effective circles of liability are in the first case less than the distance the bomb may be ordinarily thrown (and thus it may be used without thought of self-protection) and in the latter case greater, necessitating cover for the grenadier if he is to use the weapon safely.

The distinctive 'pineapple' form of the Mills bomb (and of its

American equivalent, the Mk 2) was the result of its designers trying to ensure that the case would rupture into fragments of the optimum size to kill anything in the blast zone. In fact this seldom occurred, as most bombs burst into two or three large pieces, resulting in very patchy fragment distribution patterns. It was only later that research revealed that incising the outside of the grenade casing actually had very little effect on the way it broke up, and that it is necessary to incise the inner surface to achieve the desired result, but by that time another solution to the problem had come along.

Modern 'fragmentation' grenades have bodies made either from plastic or light sheet metal, and rely for their effect not on the casing rupturing to produce lethal fragments, but on the dispersal of either hundreds of steel balls or fragments of wire, pre-notched so as to break up predictably and wrapped around the explosive core.

As well as more efficient percussion grenades, the post-Second World War period also saw the wider usage of potentially non-lethal smoke, incendiary and gas bombs, both hand-thrown and projected from launchers. White phosphorus (WP, often known as 'Willy Pete') has become the agent of choice for combined smoke/incendiary grenades. As its smoke is hotter than other types, it produces more effective cover even against infrared vision devices, while the phosphorus itself, which ignites on contact with the water vapour present in the air (and which is thus immune to attempts to extinguish it with water), is a very formidable incendiary indeed. Other types of smoke grenade produce clouds of smoke in a variety of different colours over a longer period (and thus can be used for very basic signalling – to show the limits of a position to incoming ground-attack aircraft, for example), while other types of incendiary bombs, developments of the thermite grenades used as long ago as the First World War, produce very intense heat over a longer period and are thus well adapted to the destruction of vehicles.

Lachrymogens – 'tear gas' – were first used during the Great War, too, though without any real measure of success. The development of stronger non-lethal agents such as CS, which also induces vomiting, have given them a new lease of life, particularly in such fringe military areas as crowd/riot control, and they now form a regular part of the infantryman's armoury.

The other main area of development in the two decades following the end of the Second World War saw the introduction of automatic assault rifles to replace the older semi-automatic and bolt-action weapons, together with a reduction in calibre of roughly thirty per cent. The overall effect was not only to increase the individual infantryman's personal firepower, but also to reduce the weight of his personal equipment, as the new smaller weapons make use of much lighter plastic instead of wooden 'furniture'. Assault rifles, such as the M16 and the intermediate AK47 with which the Vietnam War was fought, and the later British SA80, the French FA MAS and the Austrian Steyr AUG, all of which are much-shortened 'bullpup' designs, reverse the usual arrangement and place the magazine behind, rather than ahead of, the trigger group, and have largely taken over from sub-machine-guns. They offer the higher penetrating power and much greater range and accuracy of a full-size round while losing little of the SMG's handiness. Surprisingly, most still come equipped with bayonets, though these are now rather more likely to be multi-use implements, rather than simple stabbing blades; that of the SA80, for example, doubles as a wire cutter when employed with its sheath.

The war in Vietnam, though it was something of an anomaly, was the first where the new types of weapon predominated. All-out war, like that which had been waged in Korea, and which the Israelis were fighting against the surrounding Arab states even as the American build-up in South-East Asia got under way, it was not; in reality, it was closer in character to the 'anti-terrorist' campaigns the British had waged – and won – not far away in Malaya and Borneo, though on an infinitely larger scale. But it certainly showed the world just how much destructive power could be brought to bear by even a small force.

Open – i.e., conventional, not guerrilla – warfare, by that time, had already become an almost entirely mechanized affair, fought with tanks, artillery and aircraft, with the infantry in a subsidiary role; the Israelis demonstrated that most forcefully in 1967 and again in 1973, though the most effective manifestation was the United Nations' operation to expel the Iraqis from Kuwait in 1991. One inevitable result of that was to see close-quarter combat limited to the activities of commando units and raiding parties (and, as we shall see, to combating directly the activities of a new generation of so-called 'urban' terrorists). There are exceptions, of course – the earlier war in the Persian Gulf, between Iraq and Iran, saw battles employing tactics which were from a decidedly earlier age, and which produced staggering numbers of dead and wounded as a result. What started out as a simple attempt by Iraqi dictator

Saddam Hussein to grab 'disputed' territory from a neighbouring state soon took on a new complexion when the Iranian religious leader, Ayatollah Khomeini, declared it a jihad or Holy War, between the Muslim sects of Sunni and Shi'a. Iranian Revolutionary Guards, some of them boys as young as nine years old, were sent against entrenched Iraqi infantry in solid waves, and even voluntarily marched through known minefields in order to clear a safe path for armoured vehicles, secure in the knowledge that a martyr would go straight to Heaven. Thanks to frankly (some would say, wildly) incompetent leadership on both sides, many battles – and there were many battles – saw thousands killed for minuscule gains, or even none at all. After four years of fighting, in which perhaps 200,000 Iranians and 60,000 Iraqis died, Iraq held 700 square kilometres of what had been Iran, and Iran held a similar amount of what had been Iraq. The real victors were probably the world's tank manufacturers – a total of about 1,400 were destroyed in all.

Just as the Iran-Iraq War was getting comfortably into its stride, in early 1982, a very different sort of war started half a world away, when Argentina's ruling military junta decided to invade the Falkland Islands, ostensibly to reinforce its claim to sovereignty by force, but in reality as much to divert attention in its own country away from considerable unrest at its totalitarian practices. The result – and it came as a great surprise to Buenos Aires, which had reckoned without the British leader's iron determination that her will should be done, no matter what the cost (not to mention her own need to reinforce a shaky popularity) – was Operation Corporate, the despatch of a British combined-services Task Force to retake the islands. Without the support of heavy weapons and armour (over the efficiency of which in the islands' terrain a distinct question mark hung anyway, though the few Scorpion light tanks which did go south performed well) the battles which followed relied on distinctly old-fashioned infantry tactics, a skill which the British Army had, wisely as it turned out, never entirely neglected.

Aggression, firepower and physical fitness were the key to the success of the British Paras in their brief but violent contacts with the Argentine defenders in the Falklands. The average front-line paratroopers were nineteen or twenty years old. They had been moulded into tough and resourceful professional combat soldiers in the Parachute Regiment's rigorous twenty-two-week-long selection and training process. This seemed to have been purpose-built to prepare paratroopers for combat in the Falklands, with endurance marches carrying full combat equipment building up their fitness and endurance, and boxing (called 'milling') sessions to bring out their fighting spirit. Physically weak and poorly motivated recruits soon fall by the wayside. Parachute training has a reputation for brutality, but one Falklands veteran said the training kept him alive in combat. The 'humanely' trained Argentine conscripts, on the other hand, were like lambs being led to the slaughter, he said.

As they advanced to combat across the Falklands, British paratroopers were loaded down with weaponry. The personal weapon was the 7.62mm L1A1 Self Loading Rifle (SLR), which was the British version of the famous Belgian FN FAL. Its effective battlefield range was 600 metres and when fitted with an Improved Weapon Sight it could be fired accurately at night up to around 300 metres. Five 20-round magazines were carried in webbing pouches and several hundred more rounds were usually stuffed into pockets or a rucksack. The weapon was completed by a bayonet for close-quarter killing. Four L2 fragmentation grenades were usually carried, and a good grenadier could throw them up to thirty-five metres. Two white or colour smoke grenades were carried to create smoke screens or mark targets for artillery, mortar or close air support. Each paratrooper also carried a 66mm Light Anti-tank Weapon (LAW), which could punch through 300 millimetres of tank armour at up to 300 metres. This throwaway rocket could be quickly brought into action by extending the launch tube. It was easy to aim, having a simple flip-up sight system.

While the paratrooper's weapons were deadly and easy to use, his personal equipment and uniform were not so able to stand up to the rigours of the unforgiving Falklands climate and waterlogged terrain. The British soldier's boot was designed in the 1950s and easily let in water, causing trench foot and other problems. Sleeping bags and uniforms were equally poorly designed, so the British paratrooper was constantly fighting the elements to survive. At this point their training and physical toughness took over.

The British paratrooper fought in an eight-man section, broken down into a six-man rifle group and a two-man gun group. The rifle group was commanded by the section corporal, while the gun group was run by a lance-corporal. The gun group provided the section's firepower with a 7.62mm General Purpose Machine-Gun.

Section battle drills were simple but flexible to allow the section commander to fight according to the terrain and enemy. When the section came under effective fire, the gun group would put down suppressive fire to win the firefight. The section commander then had to plan his assault, either flanking the enemy or charging them frontally. Under covering fire, the rifle group would approach the enemy and then fight through the position in pairs. The 'battle pair' was fundamental to success. Two paratroopers would close on the enemy and assault individual trenches. One paratrooper would provide covering fire with his SLR or 66mm LAW, while the other crawled up and posted a grenade in the enemy trench and then cleared out any enemy with bullets or bayonet.

The first battle of the Falklands War, fought at Goose Green and Darwin, following unopposed landings (though Argentine aircraft sank two Royal Navy frigates during the course of them) at and around Port San Carlos, on the western side of East Falkland, was almost an infantry-only affair, though the men of the 2nd Battalion, Parachute Regiment, which fought it were partially supported by 105mm guns and those of naval assets off-shore, as well as by occasional air-strikes. They advanced on the Argentine defensive positions across largely open terrain, using fieldcraft and fire and movement, and fought through them with rifle, bayonet and grenade under covering fire from their own machine-guns and mortars, some of it rather too close for comfort. And if it was not quite a textbook operation, it succeeded none the less. Lieutenant Clive Chapman commanded B Company's 6 Platoon; it was three o'clock in the morning, local time, on 28 May:

Our platoon skirmish was initiated by the 'scarecrow' incident. As we were advancing, with two sections up, someone said, 'Watch out, there's a scarecrow in front' . . . The message was passed around the platoon and then someone else said, 'The scarecrow's moving.' The scarecrow was in fact an Argentine soldier who came towards us and said, '*Por favor*' [please]. What he actually [wanted] I don't know, but he was wearing a poncho and I said, 'Shoot him.' Private Finch, one of the GPMG gunners, opened up on the Argentinian and put quite a big burst into him. The enemy soldier literally flew through the air and the tracer ignited as it passed through him. I have no regrets . . . as it was then that we executed a platoon frontal attack upon a series of enemy trenches behind the scarecrow.

I remember thinking that the fieldcraft of the troops as they attacked was fantastic. They were weaving left and right, covering the bounds of moving men and were really getting stuck in to the fight. There was little need to hit the ground and people generally knelt to fire between moves. It was a very dark night, and we had closed to twenty to forty metres from the Argentinians when the fight began.

As to the complete number of trenches, and how many enemy were killed, I do not really know. Estimates have ranged from eight to well over twenty. Just about every trench encountered was grenaded ... Kirkwood [Chapman's radio operator] and I even took out a trench [ourselves]. There was a continuous momentum throughout the attack and it was very swiftly executed ... The Argentinian resistance was pretty weak. A lot of them were, I believe, trying to hide in the bottom of their trenches and avoid the fight ...

The success of the attack had an electrifying impact on the platoon. I think we believed from there on in that we were invincible. I am a great believer in the force of 'will' in battle, and the fact that we had imposed our will so well and so early on made us a better platoon.[1]

His company commander, Major John Crosland (known as 'The Black Hat', because he wore a black woollen cap-comforter, both on exercise and now into battle) puts it quite simply: 'The GPMGs suppressed a trench, a grenade followed, and on they went. There was to be no reorganisation as such – once the momentum was started, we were going to keep them rolling back.'[2]

The Paras met stiffer resistance as they got closer to Goose Green itself, and the advance began to slow, particularly on the left flank, where A Company, under Major Farrar-Hockley (the son of the Glosters' adjutant at Imjin) had had to attack Burntside House on its way from the startline proper, and now came up against further opposition on the low ridge east of the small settlement of Darwin, losing men (including two officers) in the assault. Soon, the battalion's commanding officer, Lieutenant-Colonel 'H' Jones, concluded that a little 'leadership from the front' was needed to get it going again, and with his headquarters ('Tac 1') he set off to supply just that. What followed is one of the best-known single incidents from the entire war, for Jones was posthumously awarded the Victoria Cross, one of two from the campaign, dying in the act of assaulting a trench system himself. (The other VC also went to a Para, Sergeant Ian McKay of the 3rd Battalion. He, too, died while attacking an Argentine position, a machine-gun post on Mount Longdon. We shall return to him later.)

Jones's decision to go into the attack himself took his headquarters staff somewhat by surprise, and as a result he had about a twenty-five-metre lead on them over the ground as he ran in a vaguely semi-circular path around the spur Farrar-Hockley's men had died trying to take, and started up the re-entrant on its far side. His bodyguard, Sergeant Barry Norman, set off in pursuit. Heading uphill now, Jones heard one of his party yell, 'Watch out, there's a trench to the left!' and turned that way. He came under fire and returned it, still running forward, with his Sterling SMG, while Norman, who had hit the ground when he heard the warning shout, emptied his SLR into the same trench, and then had a hard time changing magazines, for the next full one was jammed in a pouch on which he was lying. Jones's move to the left, Norman realized, had left him exposed to a trench on the right-hand side of the re-entrant, a danger Jones had either not seen or had chosen to ignore. Norman shouted to him to watch his back, but he paid no heed, and continued to fire into the trench on the left. A burst of automatic fire from the trench on the right caught him in the back and buttocks, and he went down. Thirty minutes later, while awaiting a helicopter to evacuate him, he died, though there is little doubt that his wound was mortal and that even immediate attention could not have saved him. One of the headquarters party, Sergeant Blackburn, said: 'It was death before dishonour stuff; but it wouldn't have passed junior Brecon' (a reference to a course in section-level leadership).

The battalion fought its way forward, platoon by platoon and section by section, in the gathering light of that early-winter morning, meeting heavier and heavier opposition the closer they got to Goose Green itself. They took heavier casualties as a result, including three who died because of a misunderstanding while trying to accept the surrender of a group of Argentinians, an incident that saw the Argentinians in question cut down by sustained machine-gun fire at close range. Sergeant John Meredith, who was firing one of the guns in question, said: 'The thing was to kill them as fast as we could. It was just whack, whack, and the more I knocked down the easier it became, the easier the feeling was. I was paying them back.' The fight for the settlement's school, close by the site of that incident, was as tough – and as confused, that other hallmark of close-quarter battle – as any, and was the closest thing to urban combat in the whole war. It was past midday by this time; the Paras had been in action continuously for some nine hours, and there were many more ahead of them before the battle would be over. Sergeant-Major Greenhalgh of C Company found himself the senior ranking Para in the move to contact:

> We had no support, no smoke and when we reached the stream I turned left and stopped to organize my group and get all the 66mms [i.e., light anti-tank rockets] and two GPMGs up behind me . . . I moved the two GPMGs on to higher ground from where I could see the school, a tent and a brick building . . . On the right of the track were two outbuildings which had not been neutralized. The Argies at the school were in trenches nearby, beyond the school and to the south, as well as in some of the buildings.
>
> I got the [GPMGs] to cover left while the 66s destroyed the two outbuildings . . . but there was no enemy [presence] in the buildings we attacked. At this stage I bumped into Corporal Harley with his gun group, and told them to cover us while we took the school.
>
> We got back to the stream and moved in single file along its bed until we came level with the school. We advanced up from the stream to a hedge line in a slight dip in the ground, about forty metres from the buildings. Then, from somewhere on our left, Captain Farrar [not to be confused with Major Farrar-Hockley] appeared with about six men . . . I told everyone to give fire support while I and two men went to the outer walls of the school; we collected all the [available] grenades before moving forward. It was an L-shaped building, with one leg pointed north towards us, with four or five windows. We lobbed L2 [fragmentation] and WP [white phosphorus, a potent incendiary] grenades through the windows and raked the rooms with our SLRs, then grouped on the right side to decide what next. There was no fire coming from the school.[3]

Said one of 11 Platoon's machine-gunners, who gave supporting fire: 'We were firing with the machine-gun from the flank and Corporal Harley ran forward with another man and each put a phosphorus grenade into the nearest building. I don't know whether there was anyone in there, but there wouldn't have been after that machine-gun fire and those grenades. We used a fair bit of ammunition; I think we put 300 rounds into it. We went on like that, clearing the buildings one by one until we reached the school.[4]

'I suppose,' Captain Farrar recalls, 'being honest, the schoolhouse fight was a free-for-all':

> ... A lot of ammunition was being discharged into the building. Bits were flying off everywhere. I had the distinct impression that the Argentinians were firing into it with heavy-calibre weapons of their own [it was the 37mm AA guns from Goose Green; they proved very effective in the point-defence role until they were neutralized, at least one with a Milan anti-tank guided missile]. I vividly recall someone firing a 66mm rocket from the prone position. The back-blast went very close to my legs. White phosphorus grenades were used [too] and the building was soon alight. The assault backed off, and I recall engaging fleeing enemy who ran off along the shoreline.[5]

Greenhalgh and Farrar then began putting down harassing fire on to Goose Green itself at close to maximum effective rifle range (about 600 metres in this case); Greenhalgh tried to involve the artillery, too, but his request for a fire mission was turned down (there were still civilians in the settlement, held hostage against just such an eventuality). All they achieved was to attract the attention of a sniper, who put down such effective fire on the shell-scrapes they were lying in that they spent the rest of the afternoon pinned down, and it was only with the coming of darkness that they could safely extricate themselves.

On the broader front, by mid-afternoon it had began to look as if too much had been asked of 2 Para, and that they would have to be reinforced, though in the event they did prove equal to the task and finally forced an Argentinian surrender. Major Chris Keeble, who took over from Lieutenant-Colonel Jones, presented the Argentinian commander, Lieutenant-Colonel Italo Piaggi, with a stark alternative to surrender with honour intact or face air-strikes and artillery bombardment, the presence of friendly civilians notwithstanding. The Argentinians, from the 12th Regiment of Infantry, and all of them recruited in Corrientes Province, surrendered just after first light on 29 May. Meanwhile, the 3rd Battalion and the Royal Marines of 45 Commando had set off on their epic cross-country march (the Paras 'tab', while the Marines 'yomp', but no matter what you call it, such a march, across truly awful terrain and

251

through icy rivers, and with each man carrying a hundred pounds and more on his back, certainly deserves adjectives like 'epic') from San Carlos via Teal Inlet towards the low mountains to the west of Port Stanley. There, they and the men of 42 Commando and the 2nd Battalion, Scots Guards, fought the battles of Mount Longdon, Two Sisters, Mount Harriet and Tumbledown Mountain respectively, while 2 Para, recovered after Goose Green and transported to Fitzroy in a spirited dash by helicopter, also fought at Wireless Ridge, the only unit to fight two major actions. The fighting on Mount Longdon, where 3 Para's men were caught, initially, in a series of steep-sided rocky gullies, was typical of the final battles.

'[The grenades] were just bouncing down the side of the rock face,' remembered Corporal Ian Bailey, leading one of the sections in 5 Platoon:

We thought they were rocks falling, until the first one exploded . . . The small-arms fire followed soon after . . . Because we had got funnelled, we weren't really working by sections now; the nearest private soldier to you just stuck with you . . .

My men started firing their 66s. Whoever was in the best position to spot targets fired; the others passed spare rockets to them. It was a very good [anti-]bunker weapon; there wasn't going to be a lot left of you if your bunker or sangar was hit by one of those. We could see their positions by now, up above us, possibly thirty feet or so away, we could even see them moving, dark shapes. Their fire was sparse to start with but then it intensified. Some of them were very disciplined, moving back into cover, then coming out again and firing again or throwing grenades.

The next cover to get forward to was in some rocks with one of their positions in the middle of it. Corporal McLaughlin's GPMG gave us cover and we put about four grenades and an 84 round [the Carl Gustav 84mm anti-tank rocket – a much heavier weapon than the 66mm LAWs] into the position, which was a trench with a stone wall around it, and a tent, which was blown over. We went round and on, myself and whichever 'toms' [private soldiers] were available, and the two men with the 84 launcher. It was over very quickly. We ran across, firing at the same time. Just as we went round the corner, we found one Argentinian just a few feet away. Private Meredith and I both fired with our rifles and killed him. The rest of the post – two men – were already dead, killed by grenades or the 84 shrapnel. We put more rounds into them, to make sure they were dead and weren't going anywhere; that was normal practice.[6]

The next move clearly wasn't so obvious. Bailey and Sergeant Ian McKay discussed the situation (if we can use such a term for what must have been a very hurried interchange indeed, on a more-or-less open hillside, under fire) and decided to try for the next cover – another Argentinian position, but this time some thirty-five metres away and centred on a machine-gun post:

He [i.e., McKay] shouted out to the other corporals to give covering fire, three machine-guns altogether, then we – Sergeant McKay, myself and three private soldiers to the left of us – set off. As we were moving across the open ground, two of the privates were killed by rifle or machine-gun fire almost at once; the other got across and into cover. We grenaded the first position and went past it without stopping, just firing into it, and that's when I got shot from one of the other positions which was about ten feet away. I got hit in the hip and went down. Sergeant McKay was still going on to the next position but there was no one else with him. The last I saw, he was just going on, running towards the remaining positions in that group.

I was lying on my back and listening to men calling to one another. They were trying to find out what was happening, but when they called out to Sergeant McKay, there was no reply. I got shot again soon after that, by bullets in the neck and hand.[7]

McKay pressed home his lone attack, and put out the machine-gun post which was the main local obstacle with grenades, but was killed in the attempt. As we have previously noted, he was awarded the Victoria Cross; for Bailey, who survived, there was a Military Medal. The 3rd Battalion lost eighteen men taking Mount Longdon, and had another thirty-five wounded; it was the costliest battle of the war for the British. Ian McKay was the oldest of the dead, and he was just twenty-nine; seven others were in their teens, and the two youngest, Private Ian Scrivens and Private Jason Burt, were just seventeen years old.

Not all the actions of that night of 11/12 June were so costly, by any means, even though one of them – the assault on Two Sisters, just to the south – culminated in one of the most bizarre adventures

253

of the war – Dytor's Charge, as one eminent commentator has dubbed it.

'It came to a point where I realized it was a stalemate,' said Lieutenant Clive Dytor, who commanded 7 Troop (a Royal Marines Troop is equivalent in size and firepower to an infantry platoon), Z Company, 45 Commando:

> and I actually remembered, at that point, a piece from a book I had read once – a book called *The Sharp End* – a bit about the Black Watch in the Second World War. They were pinned down; they had gone to ground and wouldn't get up. The adjutant had got up and waved his stick and said, 'Is this the Black Watch?' and been killed immediately, but the whole unit had got on then, surged forward. I remember thinking about that and then, before I knew it, I was up and running forward in the gap between my two forward sections. I shouted, 'Forward, everybody!' I was shouting 'Zulu! Zulu! Zulu! for Z Company.' I talked to my blokes afterwards; they were amazed. One of them told me he had shouted out to me, 'Get your fucking head down, you stupid bastard!' I ran on, firing my rifle one-handed from my hip and I heard, behind me, my troop getting up and coming forward, also firing. The voice I remember most clearly was that of Corporal Hunt, who later got a Military Medal. I think what happened was that Corporal Hunt was the first man to follow me, his section followed him, the other sections followed them, and the Troop Sergeant came up at the rear, kicking everybody's arse.
>
> So 4, 5 and 6 Sections came up abreast, pepperpotting [that is, fire-teams or sections alternately firing and moving forward in short bounds] properly. I could hear the section commanders calling 'Section up! Section down!' It worked fantastically; it was all done by the section commanders and the Troop Sergeant at the rear shouting to keep everybody on the move and the harebrained troop commander out at the front.
>
> That assault up the hill was the biggest thrill of my life. Even today [i.e., ten years later] I think of it as a divine miracle that we went up, 400 metres I think it was, and never had a bad casualty. Only one man was hurt in the troop, with grenade splinters from a grenade thrown by a man in his own section.[8]

Others heard the men themselves shouting 'Commandos! Royal Marine Commandos!' as they charged up the hill. Marine Steve

Oyitch said: 'That was to let the Argies know who was going to go in and kill them. If they chose to mix with the best in the world, they were going to get burned.'⁹ Not surprisingly, Clive Dytor won the Military Cross for his almost incredible action, and 45 Commando took one of the Argentinians' main defensive positions at a cost of just four men killed and ten more injured.

As well as Third Commando Brigade (which included the Parachute Regiment) and Fifth Infantry Brigade, the British Army also despatched components of what was rapidly becoming its star regiment, 22 Special Air Service, to the Falklands. The only significant commando action during the entire war was a raid the regiment's D Squadron carried out on an airfield on Pebble Island, off the north coast of West Falkland. On the night of 14 May, a small party from the squadron's Boat Troop, which had landed earlier, guided in helicopters carrying forty-six officers and men of the Mountain Troop while the destroyer HMS *Glamorgan* stood offshore and provided fire support with her radar-controlled 114mm gun, capable of twenty rounds a minute. The original intention had been to eliminate both the garrison (which was later discovered to have been 114 men strong) and the Pucara ground attack aircraft stationed there, but an oversight by planners meant that time was short, and the garrison escaped. All the aircraft, however (eleven of them), were destroyed in a mission which harked back to the very first SAS missions in the Western Desert. So accurate and successful was the supporting bombardment that few Argentinians ventured out of their slit-trenches and the only resistance encountered was a feeble attempt at a counter-attack, which soon petered out. All the SAS troopers returned safely to their base aboard ship, just two of them slightly wounded.

Unfortunately, many members of that successful raiding party were to die five days later when a Sea King helicopter went into the icy sea while engaged in a cross-decking operation between the carrier HMS *Hermes* and the assault ship *Intrepid*, killing a total of twenty men. The regiment had earlier been responsible for the first British offensive of the war, the retaking of South Georgia, a hazardous enterprise, but more as a result of weather conditions than through enemy action. A raid in some strength to support 2 Para at Wireless Ridge was abandoned when the four Royal Marine Rigid Raider boats the men were travelling in came under fire in Stanley harbour. Other operations 22 SAS undertook were more conventional – screening other units' activities, reconnaissance and forward observation, jobs very similar in nature to those they would undertake almost a decade later in the Middle East during Operation Granby. There has been some speculation that some of the forward observation SAS teams undertook was very forward indeed – as far forward as the mainland of Argentina – but official comment on that possibility was not forthcoming.

The Falklands war was an unusual experience for the British Army. It had been preoccupied since 1969 with fighting a long counter-terrorist campaign in Northern Ireland. This was Europe's longest running and most bloody terrorist conflict, requiring special tactics and training.

Britain's long military campaign to defeat terrorism in Northern Ireland since religious strife burst into flames in the summer of 1969 has forced the British Army to train its soldiers specially to meet the unique Irish terrorist threat.

Before British soldiers are sent out on to the streets of Northern Ireland, they undergo several months of intense training in patrolling, intelligence gathering and crowd-control skills. On top of this their shooting, fitness, first aid and basic military skills are brought up to a very high level by refresher training.

Under the direction of the Northern Ireland Training and Advisory Team (NITAT), soldiers are put through their paces in essential counter-terrorist skills. The first phase consists of training in urban skills at a purpose-built urban combat range which looks like a typical Northern Irish city. Electronic pop-up targets test troops' reactions to terrorist fire from unexpected directions and teach them how to differentiate between terrorists and unarmed civilians. The training then moves on to responding to simulated riots. In the next phase the troops have to operate in a rural environment, setting up vehicle checkpoints, manning covert observation posts, patrolling in booby-trapped areas and responding to terrorist ambushes. Airborne quick reaction forces, or 'Eagle Patrols', play an important part in this training.

Every soldier has to go through NITAT training before deployment on an operational tour. To increase the protection of its soldiers, the British Army now issues all Northern Ireland-bound troops with the Improved Northern Ireland Body Armour (INIBA), which has ceramic breast and back plates, helmet visors, 40mm Luchaire Close Light Assault Weapon (CLAW) grenade to stop armoured terrorist vehicles, and a baton round (rubber bullet) gun for non-lethal crowd control.

Once on the ground in Northern Ireland, the basic tactical unit is the four-man fire team, which is sometimes called a 'multiple' or 'brick'. This is the building block for all counter-terrorist operations. A multiple, commanded by a corporal or lance-corporal, will work together all the time to build up teamwork so they can respond instantaneously to incidents.

A typical patrol will be launched from a security force base, either on foot, vehicle or by helicopter. Vehicles are not used in so-called 'hard' border regions, where the threat of mines is great. Foot patrols are carried out by multiples in specific areas to deter terrorist activity or pick up intelligence on their movements. Usually more than one patrol is sent out from a base at a time to allow a rapid response to an incident or attack.

If multiples take fire they can only return fire if they can positively identify the firing point. This is difficult in urban areas where there is noise from traffic or factories. Finding fire positions with good cover is equally problematic, so patrols have to plan their movements with care, looking for cover as they move to prevent all the multiple being in the open at any one time. After the firing point is located and engaged, the patrol commander must call for help to try to block the terrorists' escape route. Only by quick reactions will a neighbouring multiple or airborne Eagle Patrol be able to swing into action and catch the terrorists red-handed.

The British Army's adversary in Northern Ireland is a tenacious foe. For more than twenty-five years, the Provisional Irish Republican Army (PIRA) has been fighting a vicious war against the Security Forces in Northern Ireland, the Irish Republic and on the British mainland. It is now one of the most highly skilled and best equipped terrorist groups in the world.

The 1,700 or so members of the PIRA are drawn almost exclusively from the Roman Catholic population of Northern Ireland, although there are strong contingents of southern Irish and Scottish Catholics in its ranks. Civil strife and religious bigotry stretching back centuries have generated a cesspool of discontent in Northern Ireland, which the PIRA is adept at tapping. There is still a steady stream of young recruits into the organization, who progressively gain experience and rise through its ranks. Promotion comes through bravery and success on operations.

The typical PIRA 'volunteer' joins up for a number of reasons, ranging from romantic Irish nationalism to reacting to the death or injury of family members at the hands of Protestant 'Loyalist' terrorists or their arrest by the Security Forces. Young PIRA volunteers begin their service as scouts, lookouts or couriers. They have to prove their loyalty and efficiency in these minor roles, helping more experienced terrorists set up attacks on the Security Forces.

The next progression is to become a fully fledged gunman or bomber. The volunteer will head to southern Ireland or the Middle East under an assumed identity. His relatives will have to formulate a cover story so the authorities do not take too much interest in his disappearance from their patch. At remote training bases the volunteer will be taught how to shoot and maintain the AK-47 and Armalite assault rifles which are the PIRA's favourite weapons. Particularly good shots will be taught to use the 12.7mm Barrett 'Light Fifty' M82A1, which can blast through the ceramic plates of the British Army's INIBA at up to 1,380 metres. Bomb-making is a particular speciality of the PIRA, who have created a wide variety of improvised bombs, mines and mortars from home-made explosives or the infamous Czech Semtex plastic explosive.

When sent on 'active service', PIRA volunteers use the ambush as their main tactic. They must surprise Security Force patrols and strike fast before follow-up operations can be launched to catch them. Operations range from one- or two-man sniper shoots to large-scale ambushes involving up to thirty men supported by truck-mounted heavy 12.7mm DShK or M60 machine-guns. They have to stalk their prey for days and weeks, watching for routines and weaknesses. Good covert firing positions with secure escape routes are considered essential by the PIRA. Security Force units that are lax will soon be the victims of unforgiving ambushes, either in the form of booby traps or sniper shots. A good terrorist sniper will aim to make a kill and be on the run before his victims know what has hit them.

In the Middle East, meanwhile, the Israeli-Palestinian conflict took on a terrorist character in the late 1960s, and the Israelis formed their own special forces units to counter the Arab terror threat as it evolved.

Israel's special forces units are cloaked in secrecy, none more so than the famous *Sayeret Matkal*. This commando unit is the premier long-range recce and commando unit in the Israeli Defence Forces (IDF), reporting direct to the IDF's Chief of Staff. Each of the three IDF territorial commands and each brigade has its own

Sayeret (reconnaissance) units, which contain their best soldiers. The Sayeret Matkal is the best of the best. In wartime its job is to penetrate deep behind enemy lines on strategic commando raids. Since the late 1960s it has also been Israel's premier counter-terrorist and hostage-rescue unit. It spearheaded the 1976 Entebbe hostage-rescue mission, and in 1973 gunned down a PLO chief in the heart of Beirut. More recently, it has kidnapped the radical Hezbollah leader Sheikh Abdel Karim Odeid from his house in southern Lebanon.

To become a member of this elite unit, which is not thought to muster more than 200 men, is a long and difficult process. Potential recruits are often identified in the early days of their National Service as soldiers in IDF line units. Family links are strong, and the relatives of existing or former members are often asked to join early in their military service. Unlike the British SAS there is no formal selection process. Budding Sayeret Matkal men are asked to join, after the recruiting team hear about their potential. Selection and training is tough and unforgiving. While the Brecon Beacons in winter are the proving ground of future SAS men, the Negev Desert in summer is where Israel's elite have to prove themselves. Long navigation and survival exercises with minimal kit and water are integral to the selection process. It is not unknown for potential recruits to die of heatstroke during these exercises. In spite of these hardships, there are still plenty of recruits who want to prove themselves, mainly from the Israeli Kibbutz movement, who passionately want to succeed in the Sayeret Matkal.

The Sayeret Matkal commandos use a variety of weapons. The 5.56mm Galill assault rifle and M16 with M203 grenade launcher, 7.62mm FN MAG machine-gun and 7.62mm Springfield Armory sniper rifles are all used for conventional combat missions. During close-quarter combat, hostage-rescues and assassination missions, CAR-15 short assault rifles, Heckler & Koch MP5 sub-machine-guns and silenced pistols are all favoured.

In the early 1970s, the Sayeret Matkal led the way in developing hostage-rescue techniques. Covered by snipers, its commandos perfected the techniques for assaulting hijacked airliners. Special explosives were used to blow off doors, then stun grenades were thrown inside. Commandos then quickly stormed inside to kill the hostages with precision close-quarter pistol and assault rifle fire. As Israel sought out Arab terrorists deep inside Lebanon, the Sayeret

Matkal had to begin operating under cover. Dressed in civilian or women's clothes, they had to commandeer private cars to go about their deadly business.

These tactics were used again when the Sayeret Matkal was sent after Palestinians leading the Intifada uprising in the then occupied West Bank and Gaza Strip. For long periods the Israeli specialists would stake out their prey and then launch attacks, using fast manoeuvres in cars to spring surprise raids on the Palestinian resistance.

While Israel fought its terrorist war, the Soviet Union was continuing with its campaign in Afghanistan. Following its invasion in 1979, the towns and cities fell relatively easily to Soviet conventional units and weaponry, but over 80 per cent of the countryside remained outside Communist control throughout the ten-year conflict. The Soviets and their Afghan allies were forced to wage a counter-insurgency war. This often meant fighting the Mujahedeen on foot, and in small groups. Eight years into the Soviet Afghan adventure, the Red Army had perfected counter-guerrilla tactics using small groups of special forces troops operating deep behind enemy lines. The most feared of these special forces troops were the Spetsnaz.

Originally formed in the 1950s to attack Western nuclear missile launch sites in western Europe, the Spetsnaz were at first unsuited and untrained for the war in Afghanistan. They were mainly demolition and west European language experts – skills not in demand in the mountains of central Asia. An immediate retraining programme was put into action, and the Spetsnaz were soon ready to begin hunting down Afghan guerrilla supply columns.

Small Spetsnaz teams of up to ten men were dropped off by Mil Mi-8 helicopters deep inside rebel territory. They would then patrol for days on foot looking for the trains of mules which brought weapons to the Mujahedeen from Pakistan. Moving by night, the Spetsnaz would remain hidden from view until they were ready to strike. When a caravan appeared in the valleys below, the Spetsnaz would open fire with 7.62mm PK machine-guns, 5.45mm AKM assault rifles, 7.62mm Dragunov sniper rifles and 30mm AGS 17 grenade launchers. After dispersing the caravan and possibly taking some of its members prisoner for intelligence-gathering purposes, the Spetsnaz team would move rapidly to a helicopter pick-up point to escape the wrath of the local Mujahedeen. Small anti-personnel mines and booby traps were left behind to slow down any pursuit.

To be able to move quickly through the high mountainous terrain of Afghanistan, the Spetsnaz put a premium on physical fitness and body building. They had to carry all their weapons, ammunition, rations and water on their backs, so weaklings soon fell by the wayside. Unlike their more conventional colleagues in Soviet motor rifle units, the Spetsnaz minimized the weight they had to carry. They discarded helmets in favour of bush hats, no body armour was worn to improve mobility, and ammunition had a priority over clothes, food and sleeping bags. On top of his own ammunition, each team member also carried spare belts of PK and AGS 17 ammunition.

Spetsnaz commanders thought like the Mujahedeen. They lived like them and fought like them. Every Spetsnaz soldier had to be able to survive for days or weeks away from base. Often they wore Afghan clothes to blend into the surroundings and not attract attention from curious locals.

Ambush was the classic Spetsnaz tactic in Afghanistan, and it depended on excellent personal camouflage and fire discipline. The Spetsnaz had to get the first shots in during any contact so they could escape from their more numerous opponents. Often ambushes were sprung only a few metres from their prey. At this range the automatic killing power of the PK and AKMs meant firefights were over in minutes. Spetsnaz ambushes had to be short and violent.

Artyom Borovik was a journalist who was attached to a Russian unit for a month in June 1987. He describes a typical encounter with the Mujahedeen, or the *dukhs* as he calls them:

The sweat that starts pouring out of you already on the second hundred metres is the body's vengeance for your undisciplined Moscow life. The tightly packed knapsack pounds relentlessly against your drenched back, the water bottle dangles at your side. The barrel of your Kalashnikov is trying its best to knock out one of your teeth.

Finally, thirty-two minutes and a few seconds later, the fifty raiders, including one reporter, have managed to cover 6,000 metres.

The preparation for the ambush begins right after breakfast. It consists of fitting out with equipment, receiving ammunition and communication facilities and cleaning weapons. During the night each of us will be carrying up to sixty kilogrammes: a sub-machine-gun with rounds of ammunition for it, a bullet-proof vest, a sleeping bag, rations, two canteens of water, flares and signal rockets, and rounds for the grenade launcher. All this and more has to be packed into your knapsack.

'Come on, little buddy, let's go,' Kolya Zherelin, a twenty-five-year-old Senior Lieutenant with a face brown from dust and suntan, shouts into the intercom attached to his helmet. The fellow he's addressing, whose close-cropped head can be seen through the open hatch, presses the accelerator and our IFV [infantry fighting vehicle] rolls on to the road. All of us are riding in IFVs, moving in a south-westerly direction towards Peshawar.

The border with Pakistan is only thirty kilometres away, but in fifteen minutes or so we're going to leave our vehicles and move under cover of the night due south from the road, along the Durand Line, for twenty kilometres. Then we'll lie in ambush near the villages of Singir and Biru. Information has been received that a gang is to pass through here tonight, which then plans to consolidate their positions, and tomorrow have their target practice with our helicopters.

After about five minutes all the IFVs turn from the road and make their bumpy way northwards for some three or four hundred metres. Then we leave our vehicles, form a long line and turn in the opposite direction, walking south along a dry river bed.

Each of us crouches as low as possible as he can and then dashes across the very road which he rode in comfort just a few minutes ago.

Sweat is pouring out from under my steel helmet, which I was given for greater safety. Even a chameleon would be impressed with our camouflage. Down here we turn into grains of sand, while up in the mountains we'll be rocks.

Ahead of us in the mountains we notice red and yellow flashes of light moving towards one another. At first you think there must be a highway up there in the distance. Then you realize that's impossible. What it is, in fact, are two dukh [Mujahedeen] gangs engaged in a fierce night battle.

Soon there are hills in front forming a chain in the shape of a horseshoe. We climb up and occupy all the dominant heights, the principle being – he who climbs wins. We begin at once to build firing positions. Since there are almost no large rocks on top, we have to run down to the green dry bed of a river where there are plenty of them. Each stationary shooting point, as they're called in military terms, requires twenty-five to thirty good-sized rocks. On the map, drawn at a scale of 1 to 50, the neighbouring hill is marked with the figure 642 – this is where the left flank of the covering group is located. Its right flank occupies hill 685. Captain Kozlov with his group is combing the greenery on the dry river bottom below. The gang is expected to move along the dry river bed of the Khvar river which separates the hills from vegetation. Through a pair of night binoculars several shooting positions set up by the dukhs can be seen. It's best to avoid them, though, for usually they are mined.

I feel a tickling in my throat from the sudden change in temperature, but coughing is strictly prohibited. Nor are we allowed to speak; we can communicate only with gestures and whispers. Smoking is out of the question: even if you cover the tiny speck of light with your hands a dukh can see the weak glow with the help of a night-vision instrument. As a matter of fact, dukhs can sometimes do without binoculars. But we have seven of the instruments per group. One of them dangles from my neck. At least it makes me able to see, even if I can't really hear anything. Although its batteries are still good, the binoculars should be used sparingly. Still I look through them every minute, for Zherelin has ordered me to keep the river bed under careful observation. Through the binoculars everything is tinged light green. The moon is green and so is a distant village. The lenses magnify so powerfully that I can see two little human figures in one of its streets. I mention them to Vladik Jabbarov, my neighbour in the firing position . . .

'The dukhs!' Zherelin hisses suddenly.

A drop of sweat rolls down my spine like a tiny brook running down a mountain gorge.

I look through my binoculars and see about twenty men moving at a quick pace up the river bed some distance away. They're all armed, but I can't make out what with.

We let the gang approach to a minimum distance. The tension is extreme. Kozlov moves in to close the river bed behind them and thus locks the ring. If the dukhs rush into the greenery they'll run into Kozlov's group. If they try to slip away between the hills they'll come up against us.

Shooting breaks out down below. Dozens of single and long interrupted flashes can be seen. About ten people of the gang dash haphazardly to the right bank of the river. Several figures fall. Five or six rebels lie behind boulders. An instant later they open fire on the hills, covering those trying to fight their way between our hill and a neighbouring one. On the left and right of me the sub-machine-guns of our unit are spewing bullets. The guns are aimed at three dukhs trying to outflank us on the left.

The resounding clatter tears apart the night's silence. Tracer bullets criss-cross the darkness. Several incendiary bullets hit some dry brush growth to the left of Zherelin's firing position, which instantly bursts into flames. Only the radio man is left there, for Zherelin is dashing among other firing positions.

The shooting ceases below, on the river bed; apparently Kozlov has suppressed all the firing points.

For a while it looks like red and yellow wires have been stretched across the greenery, where three dukhs are still firing back. But soon they're also extinguished and they don't reappear.

The fighting lasted about ten minutes.

The sub-machine-guns are so hot that drops of sweat sizzle as they touch the metal. Everything seems to be as it was before. Only the moon has become paler.

Suddenly firing resumes to the left of us, where two dukhs have taken cover on the rear side of a hill. The left flank of Zherelin's group is firing back from the top of their hills. One of our men appears for a second from behind a boulder and hurls something downwards. An explosion flashes brightly. Grenade fragments clang against rocks. A few seconds later another grenade explodes in the same place, just to make sure.

For about a minute we lie silently in our firing positions.

It seems like now it's really over.

A strange thought keeps pestering me. What was I doing a minute ago when I fired my sub-machine-gun at a dukh – was I defending myself or was I attacking? Was I trying to kill him or defending my life? If I were to ask a dukh about this I doubt that I would get a clear answer, even if the dukh were alive.

In the centre of the river bed, near a smooth, fat boulder glistening in the sun, lay one of the twenty men, his knees drawn up to his chin, who intended to fire rockets at our airfield at dawn tomorrow. Whom would this man, now lying helplessly at my feet, have killed tomorrow if we hadn't killed him today? The

Afghan's eyes stared in surprise at the sky, as if he wanted to ask something but couldn't. The chest of another Afghan had Sura 48 of the Koran tattooed across it in tiny letters. 'Allah didn't help him,' I thought. A third dukh lay with his head in the gravel. As he fell his right arm had been bent awkwardly under his body; it looked a very unnatural way for him to die. His left hand clutched a sub-machine-gun; we had to pry his fingers open before we could take it away.

We collected all the captured weapons and climbed back up into the hills. The soldiers occupied the firing positions and silently removed rations from their knapsacks. Jabbarov deftly spread sweet condensed milk on biscuits and crunched away content-edly. I took several swigs from a captured water bottle that was full of strong tea. It tasted slightly salty.

In about fifteen minutes we formed a long line and marched off to meet our armour.[10]

The Soviet army that invaded Afghanistan was a force trained, organized and equipped for mechanized, combined-arms oper-ations. It found it difficult to adapt to fight a counter-insurgency war. The Russians had to learn the hard way and change their tactics accordingly.

Infantry may now be mechanized for modern conflict, but the infantry still have to dismount to fight on foot, and the potential for close-quarter combat still exists. However, when a mechanized army is fighting disorganized or lightly armed opposition, infantry can remain in their fighting vehicles and pick off the opposition in a detached manner, almost like a training exercise. During the 1991 Gulf War, for example, the forces of Iraq were totally outgunned and outmanoeuvred when United Nations units launched their ground offensive to liberate Kuwait.

Operation Desert Storm saw massed US armoured units steam-rollering over Iraqi defences to crush all resistance in a hundred hours of ground combat. Advancing side-by-side with the Abrams battle tanks were Bradley infantry fighting vehicles, packed full of assault troops. Each Bradley carried a six-man infantry squad, known as dismounts, who were to take the fight to the enemy once the tanks had broken into their main defensive positions.

US infantrymen in 1991 were all professional soldiers who had joined up at the height of the Cold War. They had previously expected to fight a defensive battle in Germany, but were now leading the biggest armoured offensive since the Second World War. In command of a Bradley and its squad was a sergeant or staff sergeant. It was his job to coordinate the firepower of the Bradley's 25mm Bushmaster cannon, 7.62mm co-axial machine-gun and TOW anti-tank missiles with the shock action of his six dismounts.

Each US soldier was provided with an impressive amount of personal protection kit, including Kevlar body armour and helmet, plus a nuclear, biological and chemical warfare (NBC or MOPP) suit. This was designed to enable the wearer to survive and fight in a contaminated environment. The personal weapon of the US infantrymen was the M16A2 automatic rifle, and each squad also boasted an M249 Minimi light machine-gun. AT-4 light anti-tank rockets were also carried in the back of each Bradley for anti-armour and bunker busting.

Iraqi defences were a formidable interlocking series of earthworks and bunkers. It would be up to the dismounts to clear determined resistance from the Iraqi trenches. Once the Abrams tanks and Bradleys had destroyed Iraqi tanks, artillery and anti-tank missiles, the American armour would pour over the enemy positions right up to the front of trench lines and bunkers.

The rear ramp of the Bradleys would drop down and the dismounts would rush out the back to take up fire positions around the vehicle. Riding in the back of the vehicles was very uncomfortable – it was cramped, full of ammunition and other equipment. The dismounts were sweltering in body armour and charcoal-backed chemical warfare protection suits. They had been driving for up to two days inside the vehicles with little sleep or exposure to daylight. On dismounting the US infantry had to readjust their eyes to sunlight, locate the enemy positions and then engage them. This was a crucial moment: if they took too long to organize themselves the enemy would be able to re-establish his defence. Once out of their vehicles, the infantry would move forward to grenade the enemy trenches, while the Bradley commanders pumped rounds into the enemy positions to stop them returning fire.

In southern Iraq, US infantry had to dismount time and time again as they swept through position after position. Fortunately the Iraqis never put up any serious resistance after they had been on the receiving end of American tank or artillery fire. The main job of the dismounts was to round up thousands of stunned prisoners.

Casualties came not from dug-in machine-guns or snipers, but unexploded mines and cluster-bomb submunitions.

Sergeant Todd Regier, 'Charlie' Company, 2nd Battalion, 16th Infantry Regiment, 1st US Infantry Division, describes a typical engagement during Operation Desert Storm:

> The night before the ground war kicked off, [Sergeant McVeigh] was saying he was scared because we were going to be part of the first wave. He was scared because we weren't going to come out of it. My platoon was detached from the rest of the company and assigned to the 'ironhorse' tank company, which was tasked with protecting those units clearing the enemy trench line. Tanks and trucks fitted with ploughs followed the Bradleys, burying the enemy dead or alive in their trenches. This created a smooth crossing point for our invasion forces and also avoided the need for us to jump out of our Bradleys and engage in close-quarter combat with the Iraqis.
>
> We did a lot of firing at dug-in positions. We swung around the front-line troops and hit anybody who was dug in. We all did a lot of shooting when we came up on an obstacle. We did a lot of raking fire. At one point, we spotted an enemy vehicle about 500 yards away. McVeigh aimed the vehicle's 25mm cannon and fired, scoring a direct hit. When we plunged towards their trench lines all the Bradleys were firing at targets with their cannon and machine-guns. Targets were engaged at ranges of several hundred yards, while Iraqi soldiers were cut down at short ranges. One Iraqi soldier came out of his bunker and McVeigh fired his cannon at him. The man's head exploded – it was over 1,100 metres and shooting a guy in the head from that distance is impressive.[11]

Waiting in the desert for the Coalition advance were some 300,000 Iraqi troops. They had been in the desert for almost six months preparing for the start of the 'mother of all battles'.

In 1991, front-line Iraqi infantrymen were mostly veterans of the eight-year Iran-Iraq War. Those who did not opt to stay in the army as professional soldiers had been released from service at the end of that war. In the autumn of 1990, as the American-led Coalition gathered its forces to evict the Iraqi invasion force from Kuwait, more and more former conscripts were recalled to the ranks to provide manpower for the front. Iraqi infantry were drawn from the lowest sections of Iraqi society, and they were treated with contempt by their officers, who came from more privileged classes. They received the worst food, equipment and weapons. Their main weapon was the 7.62mm AK-47, rather than the more modern 5.45mm AKM, which went to elite Republican Guard units.

Their main job in the months up to the start of the war was to build field defences, digging trenches, bunkers, laying barbed-wire entanglements, setting booby traps and laying minefields. This was long and hard work, which had to be carried out in the scorching sun. Building defences proved a never-ending task once one sector was complete, as further defences needed building for reserve troops or out into the desert to protect the flank of the main position in Kuwait. Typical Iraqi battalion forts were triangular in shape, so other parts of the unit provided mutual fire support if the enemy broke into one part of the position. Deep minefields were sewn outside and inside the forts to slow down and channel Coalition troops into killing areas.

When the war started, the Iraqi infantrymen had little choice but to head for their bunkers and hide from the massive Coalition air and artillery bombardment. As the weeks progressed, the Iraqi frontline infantry battalions and regiments started to crack under the pressure. Supply trucks could not get through to the front with mail from home, rations and water. Slowly the Iraqi officers started to realize their fight was a lost cause, and they began to desert their men, heading for the safety of cities in the rear. Morale was at rock bottom. Coalition psychological warfare leaflets were dropped all over the front-line, offering good treatment on surrender or death if resistance continued. Random B-52 raids and MLRS rocket strikes caused panic among Iraqi troops when they obliterated company- and battalion-sized positions in minutes. Now infantry-men started to desert or surrender. Those who were caught by members of the fanatical Republican Guard or Military Police were summarily executed. Now the minefields served to trap the front-line troops in their positions.

Those Iraqis who remained in their trenches when the Coalition attack came were poorly placed to put up any resistance. Almost all their anti-tank weapons and tanks were knocked out. First they heard the American tanks at long distance, then they saw the clouds of dust. Those brave souls who tried to put up resistance were soon dead. Anti-tank missile teams tried to open fire, but were soon silenced by American tanks. When the American armour drove over the trenches, their tanks were spraying fire down into the bunkers. Iraqis who came out firing were either shot or crushed beneath the tracks of the tanks. Armoured bulldozers were also brought up by the Americans to plough up the trench lines and bury their defenders. Resistance was futile. Most Iraqi infantry just put up their hands and surrendered in their thousands.

Of course there were snippets of more traditional close-quarter combat. Allied special forces patrols which were inserted behind Iraqi lines, for example, often had to fight their way out of trouble. In such cases the fighting was vicious, unpredictable and more reminiscent of an earlier age. The British Special Air Service (SAS) patrol codenamed 'Bravo Two Zero' was compromised almost as soon as it was inserted, and its eight members had to fight a number of close-quarter actions against superior Iraqi forces. The Patrol leader, Sergeant Andy McNab, provided vivid images of these actions in his autobiographical account of the patrol's ordeals. On one occasion he and his men were engaged by APC-mounted infantry:

> Everybody knew what had to be done. We psyched ourselves up. It's so unnatural to go forward into something like that. It's not at all what your vulnerable flesh and bone wants to do. It just wants to close its eyes and open them again much later and find that everything is fine.
>
> 'Everything OK?'
>
> Whether people actually heard further down the line didn't matter, they knew something was going to happen, and they knew the chances were that we were going to go forward and attack this force that vastly outnumbered us.
>
> Without thinking, I changed my magazine. I had no idea how many rounds I had left in it. It was still fairly heavy, I might have only fired two or three rounds out of it. I threw it down the front of my smock for later on.
>
> Stan gave the thumbs up and stepped up the fire-rate on the Minimi to initiate the move.
>
> I was on my hands and knees, looking up. I took deep breaths and then up I got and ran forward.
>
> 'Fuck it! Fuck it!'
>
> People put down a fearsome amount of covering fire. You don't fire on the move. It slows you up. All you have to do is get

forward, get down and get firing so that the others can move up. As soon as you get down on the ground your lungs are heaving and your torso is moving up and down, you're looking around for the enemy but you've got sweat in your eyes. You wipe it away, your rifle is moving up and down in your shoulder. You want to get down in a nice firing position like you do on the range, but it isn't happening that way. You're trying to calm yourself down to see what you're doing, but you want to do everything at once . . .

We found ourselves on top of the [enemy] position. Everybody who could do so had run away. A truck was blazing furiously ahead of us. A burnt-out APC smoked at the far-left extreme. Bodies were scattered over a wide area. Fifteen dead maybe, many more wounded. We disregarded them and carried on through. I felt an enormous sense of relief at getting the contact over with, but was still scared. There would be more to come. Anybody who says he's not scared is either a liar or mentally deficient.[12]

As well as the SAS, there were American elite units operating behind Iraqi lines. America's elite counter-terrorist force, 1st Special Forces Operational Detachment – Delta Force – is an unconventional unit by American standards. The US armed forces are governed by strict rules and regulations that do not fit well into the free-wheeling world of special forces.

Like Britain's SAS, Delta Force troopers are trained to fight in small groups and operate without close supervision from head-quarters. The basic Delta Force unit is a four-man team, trained to fight as a self-sufficient group behind enemy lines. Team members must be medics, shooters and snipers, explosives experts, locksmiths, radio operators, linguists, mountaineers, intelligence analysts and much more. All members are volunteers who have to go through a strenuous selection process that involves long-distance marches and navigation exercises.

The rationale for the creation of Delta was the lack of a dedicated hostage-rescue force in the US armed forces, but during the 1991 Gulf War it was called upon to carry out long-range reconnaissance and sabotage missions behind Iraqi lines. Delta's first missions were far from successful. The abortive Iranian Embassy rescue attempt was followed three years later by an ill-fated mission to storm a prison on the Caribbean island of Grenada during the 1983 US invasion, in which Delta's helicopters were driven off by heavy anti-aircraft fire. In 1989, though, Delta scored a major success in freeing an American prisoner from a jail in Panama as Operation Just Cause began.

Special operations require special firepower, so Delta troopers are trained to a high degree in the use of a wide variety of weapons. Long-range firepower is provided by specially made Remington 40XB sniper rifles which can hit nine out of ten targets at 1,000 metres. Sustained and suppressive fire up to 1,200 metres is provided by M60, Heckler & Koch HK21 and M249 FN Minimi machine-guns. M79 and M203 grenade launchers add to Delta's firepower. For close-quarter kills, Delta uses the trusty Heckler & Koch MP5, CAR15, or M16 assault rifles. M-1911A1 .45 automatic pistols are carried by every Delta trooper as a secondary weapon in case his assault rifle malfunctions.

Delta was moved into Iraq by helicopter and vehicle to locate and call down air strikes on Iraqi Scud missiles. Its aim was not to take on Iraqi troops directly but to locate the Iraqi missiles covertly and then call in air strikes. Things didn't always go to plan. There were numerous contacts with Iraqi troops, which were brief, confused affairs.

Small Delta patrols were often surrounded by Iraqi troops. The key to escape was concentrated firepower. Every member of the patrol would bring his weapon to bear on the enemy. This was usually enough to stun the Iraqis and force them to back off. While the Iraqis were tending to their dead, the Delta troopers would disappear into the night.

The conflict in Chechnya which erupted between the Russian government and Chechen fighters seeking independence from Moscow was, for the Russians, a return to the early days of the Afghan conflict. Once again mechanized Russian troops were faced with a guerrilla war, and they proved themselves even more inept than in Afghanistan. Added to this, the poor tactical awareness of Russian commanders and the low morale of the troops created a nightmare scenario for the Russian Army.

Russian shortcomings were all too visible during the Battle for Grozny in the winter of 1994–95; unable to achieve success by combat action with assaults by tank and motor rifle troops, the Federal Armed Forces resorted to destructive artillery and aerial bombardment. 'The destruction of Grozny did not take the form of flattened buildings, but that of buildings gutted by fire, with their tall, gaunt, blackened, stark skeletons reaching for the sky.'[13] After

271

Grozny, it was not long before combat operations spread out westward along the Sundzha Valley in a process of 'pacification', embracing villages and settlements such as Samashki and Serno-vodsk, towns in the centre including Shali, Argun and Gudermes, and the mountain 'auls' of Bamut, Vedeno and Shatoy to the west and south. Other milestones in this conflict include the second battle at Gudermes in December 1995, the Chechen raid on Kizliyar and subsequent untidy siege at Pervomayskoye in early January 1996.

They had no uniform, no regimental badges or rank insignia and no high command. Yet the Chechens stopped the Russian advance into their capital, Grozny, in its tracks. Since January 1995, they have waged an unrelenting guerrilla war against the Russian occupation force.

The Chechens are natural fighters who in childhood are taught to use guns and knives. In the wild Caucasus region of the former Soviet Union, tribal or clan warfare has been in the blood for centuries. It was only suppressed when the Soviets imposed their rule with an iron fist, deporting the whole Chechen nation to central Asia and Siberia during Stalin's reign. By the time the Soviet Union collapsed in 1991, the Chechens were ready to seize their chance and declare independence from Moscow. The Russian army fled the republic, leaving millions of dollars' worth of modern arms behind. Soon every able-bodied Chechen male was armed with some sort of automatic weapon: 7.62mm AK-47s, 5.45mm AKMs and AKSU-74s, 7.62mm Dragunov sniper rifles, 9mm Makarov pistols, plus hundreds of 7.62mm RK and RPK, 5.45mm RPK-74 machine-guns, 30mm AGS 17 automatic grenade launchers and rocket-propelled grenades (RPG). The belt-fed AGS-17 is a particularly vicious weapon, capable of firing up to sixty 30mm grenades a minute out to 1,730 metres.

Russia's invasion united the Chechens in a way that had not been seen before. The streets of Grozny were filled with fighters preparing to do battle with the hated Russians. Their tactics were almost non-existent, but they were brave.

Chechen clans took it upon themselves to defend certain parts of the city. Relatives and friends would bring the fighters ammunition, food and fuel to sustain their battle. Due to their fanatical bravery, the Chechens were not overawed by the Russians' tanks and firepower. They would think nothing of walking out into the middle of the street to use an RPG to knock out a lead Russian BMP armoured troop carrier in a column. Other fighters would be waiting in position in high-rise buildings around the Russian column to kill any infantrymen who dismounted to clear the ambush. If the Russians cowered inside their vehicles, then the Chechens would take them out one at a time with RPGs or hand grenades.

To take the war to the Russians, the Chechens formed elite suicide squads who moved deep into the territory of the Russian Federation to take hundreds of hostages. In the summer of 1995 and in January 1996, the Chechens scored spectacular successes with these tactics. At the Kizlyar incident, for example, the Chechen fighters held off thousands of Russian troops for days in a fortified village, before slipping off into the night to escape back to their homeland. By using walkie-talkie radios and digging deep trenches, the Chechens could survive the Russian bombardment and be ready to cut down the assault troops in a hail of bullets when they made their attacks. All this came as a nasty surprise to the Russians.

Russia's war in Chechnya started with a conventional military assault on the capital that soon degenerated into chaos when the ill-prepared and poorly motivated regular army units suffered massive casualties.

To restore the situation, the Russian high command turned to the elite intervention forces of the Interior Ministry (MVD). These elite units are manned almost exclusively by veterans of the Afghan and other conflicts. Most are in their thirties and some are even older. They had been in the front line of the bloody internal conflicts that spiralled out of control as the Soviet Union collapsed. Their job is riot control, internal security, hostage-rescue and other counter-terrorist work. They come with heavy firepower – AKSU-74 mini assault rifles and AKM 5.45mm assault weapons – body armour and their own armoured vehicles. Russian body armour is designed to provide maximum protection, and has ceramic front and back plates to cover the wearer's heart. It also has attached ammunition pouches so rifle magazines can be carried.

MVD units are small, close-knit groups that have served together for many years in numerous trouble spots in the former Soviet Union. They have slick drills for setting up checkpoints and house clearing.

The SOBR (Special Rapid Reaction Group) are among the best units in the MVD's order of battle, and they made their presence felt as soon as they entered the Chechen war zone. Unlike the untrained conscripts of the regular Russian Army, the SOBR knew what they were doing. They started to set up road blocks to intercept Chechen fighters moving around the countryside. In Grozny they even got out of their BMP tracked armoured personnel carriers to trade fire with the Chechens. The SOBR began to outflank the Chechen snipers, using ruins for cover and heavy firepower to keep the fighters' heads down.

SOBR troopers were experts in the use of the 7.62mm Dragunov sniper rifle to pick off Chechen fighters as they moved among the ruins of Grozny. This was a classic street fighting battle. The SOBR had to advance slowly, house by house, to clear out the Chechens. Once they had secured a street block, checkpoints had to be set up to stop Chechens infiltrating back into the areas that had just been cleared.

Once most of the Chechens had been driven out of Grozny in early 1995, the MVD and Russian Federal Army units settled down to garrison the largely ruined city. Patrols were then sent out into the countryside to hunt down bands of Chechens who had escaped to continue the war. The strain of continued combat and lack of support back home was beginning to show on the MVD units, though. Increasingly they were affected by the drunkenness and lack of discipline that bedevilled regular units of the Russian Army. The SOBR men were beginning to realize that they were fighting for a lost cause.

The SOBR faced their biggest test at the Pervomayskoye siege in January 1996, when a fanatical Chechen group held hundreds of Russian civilians hostage. After days of bombardment, the SOBR were sent forward to infiltrate the well-prepared Chechen positions. The climax to the siege was an all-out Russian assault, in which the SOBR advanced over open ground into the village streets, suffering heavy casualties from Chechen fire and inaccurate Russian artillery support.

Even during the road movement phase of Russian armed intervention into Chechnya, there were the tell-tale signs indicating low standards of combat readiness. 'In the group of ten in which we were travelling, two tanks broke down *en route*. There was actually nothing strange about this. The earliest mark of T-72s that were produced had already undergone two if not three major base workshop overhauls. The regimental "Urals", produced in the mid-1980s, which carry the ammunition, are well matched with the 'armour'. At every stop the drivers looked under the bonnets to see whether everything was all right. Some ran to the nearest puddle with a bucket, water was escaping from the radiator too quickly....

The difference in the experience of the crews was clearly noticeable on the road. Sometimes certain tanks almost left the highway. At first it was thought that the mechanic-drivers were daydreaming because of the monotony, but then it was explained to us that these tanks contained young soldiers who were fresh out of training. They cannot always handle a heavy vehicle on an asphalt surface covered with mud.'[14]

However, it was not so much the everyday discomforts which dispirited the Russian soldier, but the total lack of clear-cut commands, doubts about the correctness of the operation and the large number of conditions which had to be satisfied before taking combat action: 'We have to think about how best to carry out combat orders, we are saddled with thirty-three conditions, by observing them we turn ourselves into sitting ducks.'[15]

The same article provides a colourful picture of a Chechen fighter, during the battle for Grozny in the early days of the siege, depicting him as

'a gunman twenty-five years old. He is wearing a black denim jacket, black denim jeans, and pointed leather boots with the trade name 'Cossacks'; he is carrying a Kalashnikov assault rifle and has around his waist a captured silver-handled Mauser inherited from his grandfather. During the day and sometimes in the evening he is at home. He eats boiled meat, potatoes, pickled cucumbers and green peas; he plays a card game called 'Duraka' with his younger sister, tries to cadge some raspberry jam from her, and then his older brother takes him in his 'Volga' to the front line to resist the well-trained Russian Army, which is armed with all the latest equipment.'[16]

In the Battle for Grozny, it soon became abundantly clear that the Chechen fighter had not lost that courageous determination, bravery and military competence which he had displayed throughout the long bitter struggle to rid himself of Russian domination over the last 200–300 years. Perhaps the most poignant and tragic episode for the Russians, during the siege of Grozny and the attempted storming of the Presidential Palace, was the annihilation of 131 Maikop Brigade on New Year's Eve 1994. During that night the brigade had managed to capture Grozny's main railway station, but then in the course of the next twenty-four hours it was literally torn apart, rendered non-effective by Dudayev's Chechens. The

Brigade lost almost all its officers, including the commander – Colonel Ivan Savin. Out of the 26 tanks that entered Grozny, 20 were set on fire and burnt out. Out of 120 infantry fighting vehicles (BMPs) only 18 were recovered from the town. All six of the 'Tunguska' AD missile complexes were completely destroyed. Over 200 officers and men were killed or missing. Furthermore, 74 men together with the corps operations officer had been taken prisoner and dozens of their comrades' half-burnt corpses littered the square in front of the Presidential Palace and along the streets.

One of the few surviving brigade officers, Lieutentant Aleksandr Labzenko, a platoon commander in an AD artillery battalion, made the following comments in an interview:

> We found ourselves in Grozny on the evening of 30th [December 1994], we were told previously that our AD battery would be attached to 81 Samar Motor Rifle Regiment (81 Samarskiy MRR) which on the 31st must go to the Chechen capital. Two 'Tunguski' were attached to 1st MR Battalion (1 MRBn), and two 'Tunguski' to the 2nd MRBn. An additional command vehicle and six to the 3rd MRBn. An officer was allocated to command each of the ZSUs. Our ZSUs were not adaptable for the conduct of urban combat operations and in principle were not suited for this role, but even so our bosses decided to strengthen the firepower of the attacking troops.[17]

On the morning of 31 December the company commander called him [Labzenko] up on the radio, assembled the column and moved off into the town. But in the suburbs, after crossing the bridge over the River Sundzha, they began to be fired on by mortars and grenade launchers. The company halted. It appeared that it had gone to a previous location of the battalion to which it was attached. The battalion had moved forward by another route. Its main objective had been to liberate the square immediately in front of the main railway station, but on the way into the town it encountered strong resistance from detachments of Dudayevites and being somewhat surprised, stopped. Finally, both the battalion and the company columns met up with each other and then speedily dispersed through the town's streets.

The BMPs moved along in column of three abreast. 'Tunguski' covered them on the right and on the left. Each kept the opposite side of the street in its sights. But suddenly the commander decided to collect a second ZSU for the reinforcement of the 2nd MRBn, and only the ZSU of Labzenko remained with the company. Not far from the street in which the hospital was located, tanks supporting the motor riflemen began to come under Chechen grenade launcher fire from all sides – the Chechens were professionals producing very accurate fire. Immediately two of the leading tanks burst into flames, three others began to crawl along the sides of the street. The BMPs increased speed, but clashed with an advancing column of combat vehicles moving towards the palace square. 'The road is blocked' Labzenko reported to the company commander, 'what shall we do?' The company commander ordered him to use the map – 'in front it must be clear – turn left'. The day was drawing to a close and the light was failing, it was difficult to see the road from the vehicles, but they turned left, up to the second bridge over the River Sundzha. Behind him, just as the vehicles were stretched out in a side street, grenade launchers appeared once again, and two tanks were set alight – one in the lead and one at the rear. The BMPs and the 'Tunguska' had been caught in a trap. Labzenko continued, 'Luckily for us, nearby, some hundred metres or so from the town hospital, there turned out to be a vehicle service station forecourt. The whole company made a run for it under the cover of a wall.'[18]

At the very same time the main body of 131 MR Bde – its headquarters and commander, 1st and 2nd MRBns with attached enhancements, the remnants from the allotted tank battalion which had not been set on fire – occupied the main railway station. It also turned out that they had been surrounded by hundreds of Dudayev's Chechen fighters. They were perched on every level and floor of buildings adjoining the square by the main railway station, in basements, on roofs and at every window. Grenade launchers and snipers were unceasing in their activities, switching fire from one vehicle to another, putting any vehicle out of action that moved away from the cover of the wall. The internal troops, who in accordance with the plan of operations should have moved forward as the second echelon in support, had become separated from the first echelon by the Chechen encirclement of the railway station.

Testimony of a Russian Lieutenant-Colonel in 81 MRR: 'Our unit on 31 December at 08.00 hrs moved into Grozny and occupied a defensive position near the central railway station. At 14.00 hrs the first BTR was hit by a grenade launcher, and after an hour the battle began and continued for a whole twenty-four hours. During this time the Chechens destroyed fifteen tanks, and toward the evening of New Year's Day, all that remained of the reinforced battalion, which had entered Grozny, were sixty men plus forty-five wounded, only 30 per cent of the original total.'[19] Thus, 131 MR Bde ceased to be an effective fighting entity; many Russian raw recruits and conscripts paid the price for their senior officers' and commanders' complacency.

There is another aspect in the Chechen conflict which requires mention. It is the occasion when command, control and discipline break down leading to unrestrained, cruel and barbaric action against the civilian population. Such was the case at Samashki, encapsulated in the words of Aminat Gunasiyeva, a resident of Samashki:

> On the morning of April 7, the commanders said that unless we surrendered 286 sub-machine-guns to them by 4 p.m., an assault on the settlement would start. But on that day our position was hopeless. In no way could we manage to get the number of guns demanded so soon. We asked for a week. But evidently, the ultimatum was merely a pretext because nobody waited until 4 p.m. as promised. It all started two hours earlier. There we sat, unable to escape. We heard the gates being opened, the bolt being pulled out of the lock, a BTR driving in, and a hand grenade being hurled into the empty basement. They entered into the room. There were about eighteen to twenty of them. They looked sober, only their eyes were kind of glassy with an angry heat in them. They themselves seemed frightened. We have an order to kill everyone between the age of fourteen and sixty-five.[20]

In June 1995, as an extension of the 'pacification' process, the Federal Armed Forces conducted large-scale combat operations in the mountains in the south against the illegal 'bandit' formations of Dudayev. A key operation was that of the seizure of Shatoy, long regarded as a vital Chechen base by the Russian Federal Command. The concept of the operation against Shatoy was that motor rifle troops would use two routes from the north, one being through the defile of the Argun River. In the rear of the Chechen position three airborne descents would be effected to block reinforcements.[21]

245 MRR dashed, in a classical sense, headlong into an ambush. Although it was well known that the Dudayevites would expect an assault to come here, having carefully prepared for it, and having

fortified and improved it by the use of engineer enhancements. The regimental reconnaissance was deployed ahead of the 2nd MRBn. Not having noticed the Dudayevite positions on the slopes, they fell through and over the line which marked the 'fire sack'.[22] Signaller Private Viyacheslav Osin became the first victim of the leaden avalanche. Private Panfilov sprung from the armour and began to fire in short bursts of machine-gun fire, switching from one enemy fire point to another. He attracted enemy fire on himself and gave the others a chance to repulse the attack.

Following the Chechen raid by Salman Raduyev on Kizliyar on 10 January 1996, his selection of the return route via Babayurt and Pervomayskoye was not by accident. The villages of Sovetskoye and Terskoye (Terechnoye), near Pervomayskoye, were mainly inhabited by Chechens, who were ready and prepared to meet them. Raduyev had already worked out a plan, while in Kizliyar, in which it was envisaged that inhabitants of these villages would establish a human corridor for the fighters on their way back to Chechnya. However, on 10 January 'at 10.00 hrs the column with the hostages was stopped some ten kilometres from the Chechen–Daghestan administrative border, on the outskirts of the village of Pervomayskoye, situated thirty-five kilometres to the north of Khasavyurt.'[23]

Even the elite special forces were not immune from confusion and lack of clear tasking in the command chain:

For almost a week we sat without any information. Only by our own efforts did we learn that the 'Raduyevites' had not wasted any time. While we froze in an open field waiting for some kind of orders, the fighters compelled the hostages, in the first place, the OMON prisoners, to dig trenches. For this very reason the trenches were dug competently – around the perimeter and in the village itself. They used a local mechanical crane to construct real pillboxes out of concrete blocks from which the fighters fired on our troops later. Certainly, they were well supplied with captured weapons – heavy calibre machine-guns, anti-tank rockets, assault rifles, grenade launchers and their snipers had powerful optical sights. The tactics that they used were simple – the 'Raduyevtsy' fired from behind the hostages, who were positioned in front of them.[24]

As an escaped hostage there were dangers and inexplicable hostile treatment by fellow countrymen:

During the daytime the hostages were made to stand in front of the trenches in deep mud. In the evening they were herded back into buses and houses in the village. Vladimir Timoshenko took advantage of the chaos and escaped. He reached the first military outpost on all fours in the mud. The Russian soldiers beat him with a ramrod; eight times on the back and once on the head. When he fell in the mud again they began to kick him. They then stripped him naked to see if he had spoken the truth about the Chechens beating him.[25]

When the assault eventually went in, it was obvious that it was doomed from the start for:

the first wave of the 'stormers' broke away to a flank, not reaching the village boundary. Then, the Daghestani rapid reaction group pushed on ahead. It moved ahead of 'Al'fa' and received the first casualties. There was no overall communication net and coordination. Sub-unit radios worked on their own frequency. For coordination, it was necessary to yell across a field, to overcome the thunder of battle. One of the special forces sub-units during the 'storming' requested the headquarters to confirm an airstrike detailing the coordinates of a Dudayevite strong point. It bawled: 'Switch the attack over there.' But the headquarters replied, 'Sorry, we cannot communicate with helicopters.' Thus, a tragic incident, aviation 'had brought down fire' on its own [troops] – through a lack of forward air controllers (FACs).[26]

Another factor which hindered operations for the Russian Federal Forces was the severe weather.

Twilight over Pervomayskoye – a dismal picture. The grey-blue heavy sky. The fast-thickening fog. After five in the evening it had already darkened completely. Everywhere reeds, dirt, was covered in numbing snow. The temperature was 2–3 degrees of frost, but with a piercing wind. In such dampness – it is a catastrophe. A man is momentarily frozen to the bone, even if he is on the move. And if he is required to sit in an ambush or listening [post] for two hours, you begin to become numb with cold, stupid with hunger and think only about how it would be in the warm. Among the Russian soldiers were several victims with burns. In attempting to warm themselves some poured diesel [solyarka] on the camp fire – setting clothing alight.... Some suffered from shock, having put themselves too close to an engine. Actual casualties of federal forces in all consisted of not less than fifty to sixty men.[27]

During the night of 17/18 January attempts by Raduyev fighters to break out had already started. Before this, on the same evening around 23.30 hrs, from the direction of the village of Sovetskoye, fighters of Basayev hurried to help Raduyev and attacked the outside ring of Federal Troops. During the night striving to pierce the encirclement, the shock group of Raduyevites began to break out. As a result, fifty-two fighters were killed. Despite heavy casualties, a group of fighters was successful in getting away to the village of Bulat Yurt in a north-easterly direction. This village was inhabited solely by Chechens. Other fighters were ferried across the River Terek in boats, prepared beforehand by the villagers. Raduyev himself, it is thought, abandoned his people and went off alone. Some Chechen fighters escaped by running along the threadlike shape of a gas pipeline over to the far bank of the River Terek.

Events since Pervomayskoye have continued to show repeated mistakes by the Federal Forces, indicating a general lack of motivation. Whereas, on the Chechen separatists' side, despite the reported death of Dudayev, their campaign to rid their soil of Russians has continued unabated and with greater intensity.

Close-quarter combat in general is a visceral aspect of soldiering which has remained unchanged through a period when technology has taken over the battlefield and turned it into a place where the life expectancy of an unprotected man is very low indeed. The bitter civil war in the former Yugoslav republic of Bosnia-Herzegovina saw close-quarter combat take on a new character, with neighbours taking up arms against each other in a brutal fight for ethnic supremacy and separation.

At the start of the Bosnian war in 1992, the Croatian Defence Council (HVO) played a key role in stopping Serb attacks in their tracks. Within a year they had turned on their Muslim allies and were engaged in bitter urban fighting in Mostar and the towns of central Bosnia.

Many HVO commanders learnt their trade in the old Communist Yugoslav Army before the war, but most of the ordinary soldiers had no previous military experience. The Croats, however, had plenty of money to buy weapons and local knowledge of the terrain.

They were tenacious street fighters and were rarely defeated on their home territory. The HVO's biggest stronghold was Mostar in south-west Bosnia. They first fought off Serb attacks and then took the offensive to drive the local Muslims into a small ghetto in the east of the city.

HVO militiamen signed up to defend their homes. Formal military instruction was poor in the early days of the war, but by 1993 formal training camps were set up to school Croat youth in military tactics and weapons handling. The main weapon of the HVO was the Yugoslav-made Zastava M70B1/AK-47 Kalashnikov assault rifle with folding stock. Highly prized was the Zastava M76 7.92mm semi-automatic sniper rifle, which proved deadly in urban warfare. Yugoslav-made RPG-7 rocket grenades and RPG-22 72mm Light Anti-Tank Weapons provided heavy close-quarter firepower and were generally employed as counter-sniper weapons.

When setting up a defensive position, the HVO would first round-up civilians from rival ethnic groups and expel them, often at gunpoint, from the local area to prevent information being passed to the enemy. Road blocks were established to provide early warning of enemy attacks. The main defensive positions were then prepared behind massive minefields. HVO militiamen had to learn fast how to plant mines and link them into their fire positions. Every minefield had to be covered by sniper positions to break up and slow enemy attacks.

When attacks came in, the HVO quickly manned their trench lines to fight off the enemy. Their local knowledge now paid dividends. Every man was given a specific field of fire to cover. He would shoot everything that moved in front of him. Positions were reinforced with sandbags and overhead cover. In urban areas, firing positions were built inside ruined buildings to disguise the flash and smoke of their weapons. HVO defences proved formidable – the front line in Mostar remained static for more than a year.

For the armies of Western countries, the heavy casualties associated with close-quarter infantry combat have made governments and high commands very hesitant to commit their forces to action in urban areas – witness the American reluctance to press the pursuit of the defeated Iraqi army into the cities of Basra and Baghdad. The American-led UN aid mission to Somalia in 1993 saw the US military drawn into heavy street fighting for the first time since the 1989 invasion of Panama.

As the US Army's elite strike force, the Rangers are trained to carry out raids and ambushes deep behind enemy lines. In August 1993, Task Force Ranger, comprising the 3rd Battalion, 75th Ranger Regiment, supported by elements of Delta Force and the 160th Special Operations Aviation Regiment ('The Night Stalkers'), was deployed to the Somali capital Mogadishu to hunt down local warlord Mohammed Farah Aideed. American commanders of the United Nations force in Somalia wanted to capture Aideed after his militia had staged a number of deadly attacks on their peacekeepers.

The Rangers started to stage a series of helicopter raids against Aideed's suspected hide-outs. Squads of Rangers were flown out to the slums of Mogadishu in Night Stalkers UH-60 Black Hawk helicopters. When they got to the target, the Rangers fast-roped to the ground as the helicopters hovered ten metres above. Once on the ground, the Rangers were on their own in the middle of a largely unknown and hostile city.

Each Ranger squad had to be self-sufficient, and were equipped with extra ammunition, food and water in case the helicopters couldn't get back to pick them up. The eight-strong squads were commanded by a young sergeant in his mid-twenties, while most of the soldiers were aged between eighteen and twenty. M16A2 assault rifles, M249 FN Minimi light machine-guns and M203 grenade launchers were the Rangers' favoured weapons. Even though they had been highly trained at the US Army's Ranger School, the Somali operation was the first combat mission for many of the members of Task Force Ranger.

On 3 October 1993 disaster struck. A major raid went wrong when the Somalis managed to shoot down two Black Hawks. Some seventy Rangers found themselves surrounded at night in a small perimeter, with hundreds of Somali militiamen pouring AK-47 fire into their positions. Regular impacts by RPG-7 rockets added to their discomfort. The Rangers took cover in mud buildings and started to return fire. They used classic fighting in built-up area (FIBUA) tactics to hold their ground until a relief column could fight its way through. Each Ranger was given an arc of fire around the perimeter to cover with fire. If any Somali appeared in a Ranger's sight he was dead. The Minimis and M203s were used to take out snipers or machine-gunners who tried to open fire on the Rangers from high buildings. Somali tactics were unsophisticated, but they were determined. They would regularly try to rush the buildings occupied by the Rangers, only to be cut down in a hail of bullets from the Americans' automatic weapons. Strict fire discipline was the key to the success of the Rangers' defence. Every man held his fire until a target appeared, and was careful not to waste ammunition. More than twelve hours after being landed, the Rangers were able to escape when an armoured relief column punched through the Somali siege. Some eighteen Americans were killed and seventy-seven injured during the action. The Somalis paid a high price for their victory, with more than 300 dead filling the streets around the battle site.

So what, in the end, makes a successful close-quarter fighter? Training, or instinct? We have seen how seemingly shy men, conscripts rather than professional soldiers, unused to violence as a way of life, suddenly turn into killers in reaction to the violent deaths of comrades in arms, and how others take fatal risks in order to save or spare their fellows. But equally, there must always be the suspicion that would-be killers regularly join the world's armies in order to seek out the chance to murder legally and without hindrance. Soldiers are human, after all: some shrink from violence, yet embrace it at need; others seek actively for ways to indulge in it; still others recognize that there is a time and a place and even a need for anything and everything, including visiting sudden, violent death upon one's enemies, face to face and belly to belly.

It is worth recalling Dan Fairbairn's words, perhaps, written in the dark days of 1941, but with a certain wider currency, none the less:

> Some readers may be appalled at the suggestion that it should be necessary for human beings of the twentieth century to revert to the grim brutality of the Stone Age in order to be able to live . . . But it must be realized that, when dealing with an utterly ruthless enemy who had clearly expressed his intention of wiping this nation out of existence, there is no room for any scruple or compunction about the methods to be employed in preventing him. The reader is requested to imagine that he himself has been wantonly attacked by a thug . . . Let him be quite honest and realize what his feelings would be. His one, violent desire would be to do the thug the most damage – regardless of rules. In circumstances such as this he is forced back to quite primitive reactions.[28]

Sources

Chapter One   Fifty-two Months in Hell

1 Paddy Griffiths, *Battle Tactics of the Western Front* (Yale University Press, New Haven and London, 1994)
2 Lionel Crouch, *Letters from the Front* (printed for private circulation, 1917)
3 J.C. Dunn (ed.), *The War the Infantry Knew* (P.S. King, Edinburgh, 1938); (reprinted Jane's 1987 and Sphere Books 1989)
4 Ibid.
5 Henry Dundas, *Henry Dundas, Scots Guards: a Memoir* (Blackwood, 1921)
6 Dunn, op. cit.
7 Dunn, op. cit.
8 Dunn, op. cit.
9 Dunn, op. cit.
10 Dunn, op. cit.
11 Dunn, op. cit.
12 Dunn, op. cit.
13 Dunn, op. cit.
14 Dunn, op. cit.
15 'A Rifleman' [Aubrey Smith], *Four Years on the Western Front* (Odhams, London, 1922)
16 Ibid.
17 W.E. Fairbairn (with P.N. Walbridge), *All-in Fighting* (London, 1942); (reprinted by Paladin, Boulder, no date)
18 Rifleman T. Cantlon, KRRC, quoted in Lyn Macdonald, *Somme* (Papermac, London, 1984)
19 Quoted in Bryan Cooper, *The Ironclads of Cambrai* (Pan Books, London, 1967)

20 Ibid.
21 Ernst Junger, *Storm of Steel* (Chatto & Windus, 1929)

Chapter Two   A violent *entr'acte*: Rearmament and Civil War in Europe

1 Ralph Bates, 'companero Sagasta Burns a Church', in *New Left Review* 13 October 1936. Quoted in *Spanish Front* (Oxford University Press)
2 Quoted in *The Book of the XV International Brigade* (Frank Graham, Newcastle, 1975); (reprint of the original, published in Madrid, 1938)
3 Ibid.
4 Ibid.
5 Ibid.
6 Ibid.
7 Richard Meinertzhagen Diaries (unpublished), quoted in Mark Cocker, *Richard Meinertzhagen: Soldier, Scientist and Spy* (Secker & Warburg, London, 1989)
8 Ibid.

Chapter Three   Blitzkrieg!

1 Franz Kurowski, *Infantrie Aces* (Fedorowicz Publishing, Winnipeg, 1994)
2 Ibid.
3 Anthony Hechstall-Smith, *Tobruk*, quoted in Don Congdon (ed.), *Combat in World War II* (Arbor House, New York, 1958)
4 Ibid.
5 Kurowski, op. cit.
6 John Jeffris, quoted in Colin John Bruce, *War on the Ground* (Constable, London, 1995)
7 H.P. Samwell, *An Infantry Officer with the 8th Army* (Blackwood, London, 1945). Samwell died in the Ardennes Campaign in 1944.
8 Quoted in Alan Clark, *Barbarossa* (Weidenfeld & Nicolson, London, 1965)
9 Ibid.
10 Ibid.
11 Ibid.

12 Alvin M. Josephy Jr, *The Long and the Short and the Tall: the Story of a Marine Combat Unit in the Pacific* (Zenger Publishing Co. Inc., Washington, 1975)

13 William Manchester, *Goodbye Darkness* (Michael Joseph, London, 1981)

14 John Laffin, *Tommy Atkins* (Cassell, London, 1966)

15 Quoted in Charles Whiting, *44 – In Combat from Normandy to the Ardennes* (Military Heritage Press, New York, 1984)

16 Glover S. Johns, *The Clay Pigeons of St Lô* (The Military Service Publishing Co., 1958)

17 Kurowski, op. cit.

18 Peter Stainforth, *Wings of the Wind* (Falcon Press, London, 1952); (reissued in 1988 by Grafton Press and Arms & Armour Press)

19 Ibid.

20 Ibid.

21 Quoted in Cornelius Ryan, *A Bridge too Far* (Pan Books, London)

22 John Dominy, *Escapers*, quoted in Whiting, op. cit.

23 Quoted in Whiting, op. cit.

24 Laurence Critchell Bastogne, quoted in Congdon, op. cit.

25 Andrew Wilson, *Flame Thrower*, quoted in Congdon, op. cit.

26 Ibid.

Chapter Four **Commandos and Special Forces**

1 William Seymour, *British Special Forces* (Sidgwick & Jackson, London, 1985)

2 W.E. Fairbairn (with P.N. Walbridge), *All-in Fighting* (London, 1942); (reissued by Paladin, Boulder, no date)

3 Ibid.

4 Ibid.

5 Ibid.

6 W.E. Fairbairn and E.A. Sykes, *Shooting to Live with the One-hand gun* (Faber, London, 1942)

7 Ibid.

8 Ibid.

9 Quoted in Mike Langley, *Anders Lassen, VC, MC, of the SAS* (NEL, London, 1988)

10 Ibid.

11 Ibid.

12 Peter Young, *3 Commando, Vaagso, 1941* (Orbis, London, 1983)

13 Ibid.
14 C.E. Lucas Phillips, *The Greatest Raid of All* (Little, Brown & Co., New York), quoted in Congdon, op. cit.
15 Langley, op. cit.
16–38 Ibid.
39 George Iranek-Osmecki (trans.), *The Unseen and Silent* (Sheed & Ward, London, 1954)

Chapter Five Britain's Brush-fire Wars: the End of Empire

1 Charles Allen, *The Savage Wars of Peace* (Michael Joseph, London, 1990)
2 Michael Dewar, *Brush-fire Wars* (Robert Hale, London, 1990)
3 Ibid.
4 Allen, op. cit.
5–20 Ibid.
21 Michael Paul Kennedy, *Soldier 'I' SAS* (Bloomsbury, London, 1989)
22 Ibid.

Chapter Six America's Asian Wars

1 Anthony Farrar-Hockley, *The Edge of the Sword* (Frederick Muller, London, 1954)
2 Ibid.
3 Ibid.
4 Ibid.
5 Ibid.
6 S.L.A. Marshall, *Pork Chop Hill* (William Morrow, New York, 1956)
7 Ibid.
8 Ibid.
9 Robert Sanders, *Brothers* (Presidio, Novato, 1982)
10 Captain Moyers S. Shore II, *The Battle for Khe Sanh* (History and Museums Division, US Marine Corps, Washington, 1969)
11 Ibid.
12 Tom Mangold and John Penycate, *The Tunnels of Cu Chi* (Hodder & Stoughton, London, 1985
13 Ibid.
14 Ibid.

15 Ibid.
16 Ibid.

Chapter Seven **Through Tribal War to Techno-war**

1 Lieutenant Clive Chapman, quoted in Mark Adkin, *Goose Green* (Leo Cooper, London, 1992)
2 Major John Crosland, in ibid.
3 CSM Greenhalgh, in ibid.
4 L/Cpl Michael Robbins, quoted in Martin Middlebrook, *Operation Corporate* (Viking, London, 1985)
5 Captain P. Farrar, quoted in Adkin, op. cit.
6 Corporal Ian Bailey, quoted in Middlebrook, op. cit.
7 Ibid.
8 Lieutenant Clive Dytor, in ibid.
9 Marine Steve Oyitch, in ibid.
10 *Afghanistan in our lives* (Novosti Press Publishing House, Moscow, 1989)
11 *The New York Times*, Thursday, 4 May 1995
12 Andy McNab, *Bravo Two Zero* (Bantam, London, 1993)
13 Charles Blandy: an observation made during a visit to Chechnya, 2–7 December 1995
14 *Izvestiya* No. 2 (24,361) 6 January 1995, p. 2
15 Ibid.
16 Ibid.
17 *Izvestiya*, 11 January 1995, p. 4
18 Ibid.
19 N.N. Novichkov, *Rossiyskiye Vooryzhenniye Sily v Chechnskom Konflikte: Analiz – Itolgi – Vvody* (Parizha, Moscow, 1995)
20 *Moscow News* No. 37, May/June 1995: article by Aminat Gunasieva
21 *Krasnaya Svezda*, 24 June 1996
22 Killing Ground: idiomatic translation
23 *Krasnaya Svezda*, 11 January 1996
24 *Ogonek*, No. 5, January 1996
25 *Gazeta Wyborca*, No. 18, 22 January 1996
26 *Zavtra*, No. 4, 1996
27 Ibid.
28 Fairbairn, op. cit.

Index